SpringerBriefs in Physics

T0202695

More information about this series at http://www.springer.com/series/8902

Haiyin Sun

A Practical Guide
to Handling Laser
Diode Beams

 Springer

Haiyin Sun
ChemImage Corporation
Pittsburgh, PA
USA

ISSN 2191-5423   ISSN 2191-5431   (electronic)
SpringerBriefs in Physics
ISBN 978-94-017-9782-5   ISBN 978-94-017-9783-2   (eBook)
DOI 10.1007/978-94-017-9783-2

Library of Congress Control Number: 2012939640

Springer Dordrecht Heidelberg New York London

Printed on acid-free paper

Springer Science+Business Media B.V. Dordrecht is part of Springer Science+Business Media (www.springer.com)

Foreword

Since the invention of the laser in the 1960s, lasers have found extensive applications in many areas and the laser market has grown rapidly. According to a report published in January 2013 in *Laser Focus World* [1] the global laser market was worth $8.62 billion in 2013, an increase of 62 % from $5.33 billion in 2009, and the laser diode market constantly accounts for about 50 % of the dollar amount. Laser diodes are also the most widely used lasers from an application point of view. Below is an incomplete list of applications:

Alignment	Bio cell detection	CD data record and reading
Confocal microscopy	Capillary electrophoresis	Display and entertainment
DNA analysis	Flow cytometry	Genomics
Gas detection	High-speed printing	Holography
Imaging on film	Inspection	Interferometry
Laser-induced fluorescence	LIDAR	Lithography
Machine vision	Materials processing	Metrology
Particle counting	Raman spectroscopy	Rangefinding
Remote optical sensing	Reprographics	Telecommunications

Compared with other lasers, laser diodes have unique characteristics and offer a number of advantages:

1. A wide range of wavelength selection from ultraviolet (375 nm) to middle infrared (10.4 µm).
2. A wide range of power selection from mW to kW.
3. Operation either in continuous wave or pulsed mode, with pulses as short as picoseconds, or modulated up to gigahertz rates.
4. As small as a needle tip.
5. High electrical to optical power conversion (efficiency of over 30 %) and capable of battery operation.
6. Long lifetime of about 10,000 h.

However, laser diodes also have a number of shortcomings:

1. Highly divergent, elliptical and astigmatic beams.
2. Unstable wavelengths and powers.
3. Sometimes have multi-single modes and/or multi-transverse modes.
4. Large manufacturing tolerance.
5. Vulnerable to electric static discharge.

These shortcomings often make the application of laser diodes a challenging task. A good understanding of the unique properties of laser diodes, and particularly of how to manipulate and characterize diode laser beams, is essential for the effective use of laser diode systems.

Many optical design technical books have been published. These books mainly deal with imaging optics design based on geometric optics using the sequential raytracing technique. Some books touch briefly on the subject of optical design as related to laser beam manipulation. On the other hand, many books on laser diodes have been written. These books all extensively deal with laser diode physics with little or no discussion of laser diode beam manipulation or characterization. Some internet resources dealing with laser diode beam manipulation can be found online. However, in this author's opinion, these resources do not provide enough material or sufficient detail on laser diode beam manipulation and characterization.

This book intends to address this vacancy and provide a practical guide and reference to those scientists and engineers who are still new to laser diode applications, and to those undergraduate and graduate students who are studying lasers and optics. The author hopes that the readers will be able to quickly and easily find the most practical and useful information about laser diode beams from this book without having to search through a sea of information.

This book *A Practical Guide to Handling Laser Diode Beams* is a revised and significantly extended version of the book *Laser Diode Beam Basics, Manipulations and Characterizations* written by the author and published by Springer in 2012. Since the publication of the previous book, the author has received much useful feedback from readers, and was motivated to write a follow-up book to address these comments. Compared with the previous title, the book contains about 75 % more content, covers many more topics, discusses the subject matter in more detail, and has been extensively reorganized. The new topics include: laser diode types and working principles, non-paraxial Gaussian beams, Zemax modeling of Gaussian beams, numerical analysis of laser diode beam characteristics, spectral property characterizations, and power and energy characterizations. Much of the existing text has also been revised to include more detail, and many graphs have been redrawn.

Nowadays, a lot of information can be found by online keyword search. Therefore, in this book we list only some references that are rather specific and not easy to find through a general search. We also list some Internet resources. A few references are recommended for those readers who are interested in learning more about laser diodes: References [2] and [3] for laser diode physics, Rreference [4] for general questions about laser diodes, and Reference [5] for a concise review of laser diode history and applications.

References

1. Laser Market Place: Laser markets rise above global headwinds. http://www.laserfocus world.com/articles/print/volume-49/issue-01/features/laser-marketplace-2013-laser-markets-rise-above-global-headwinds.html (2013)
2. Coldren, L.A., Corzine, S.W.: Diode Lasers and Photonic Integrated Circuits. John Wiley and Sons, Inc., New York (1995)
3. Pospiech, M., Liu, S.: Laser diodes. An Introduction. http://www.matthiaspospiech.de/files/ studium/praktikum/diodelasers.pdf
4. Sam's Laser FAQ. http://www.repairfaq.org/sam/laserdio.htm
5. Welch, D.: The laser diode 50 years and counting. http://www.infinera.com/technology/files/ Infinera-Celebrating-50-years-of-Diodes.pdf

Contents

Chapter 1
Laser Diode Basics

Abstract The basic optical, electrical, and mechanical characteristics and the working principles of laser diodes are summarized. Vendors and distributors for laser diodes, laser diode modules, and laser diode optics are introduced.

Keywords Active layer · Band · Carrier · Cavity · Current · Gain · Junction · Laser diode · Lens · Linewidth · Modes · Module · Power · Spectral · Tolerance · Vendor · Wavelength

Laser diodes are unique compared with other types of lasers. A little background knowledge of laser diodes will be helpful for the readers to understand the contents of this book. We will only briefly summarize this background knowledge without involving a lot of physics and mathematics.

Because laser diodes have manufacturing tolerances larger than other types of lasers, laser diodes of the same type often behave differently, in terms of wavelength, power, threshold, beam size, beam divergence, beam pointing, etc. When we talk about parameter values for laser diodes in this book, we use terms such as "typical value" or "typical value range" because of the large tolerance.

1.1 Laser Diode Types

1.1.1 Homojunction Laser Diodes

The beam of a laser diode is generated inside an active layer. The earliest laser diodes had a p-n-type homojunction as the active layer. The p-n junction is made of the same material, as shown in Fig. 1.1. The active layer has a submicron thickness.

When an electrical current is injected through the p-n junction, the electrons are in the conduction band and the holes are in the valence band, as shown in Fig. 1.2. There is an energy gap between the two bands. Electrons at the higher energy level can recombine with the holes at the lower energy level and emit photons. The photons are mainly confined inside the active layer and can only travel inside the active layer in the z direction shown in Fig. 1.1. Some photons are absorbed by the

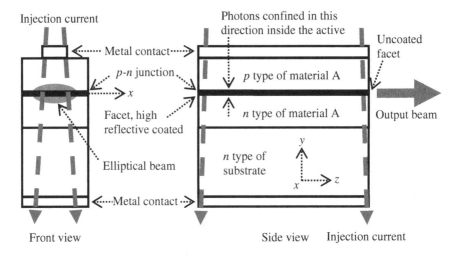

Fig. 1.1 Schematic of the front view and the side view of a homojunction laser diode

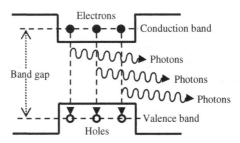

Fig. 1.2 Electrons in the conduction band recombine with the holes in the valence band to emit photons

material of the active layer while traveling, which is the loss mechanism of the laser. Some photons induce more electrons and hole recombinations and generate more photons as they are traveling inside the active layer. This is the gain mechanism of the laser. One facet of the active layer is high reflection coated to provide >99 % reflection. Another facet of the active layer is uncoated with a natural reflectivity of 30–40 %. When the photons are incident on the uncoated facet, a portion of the photons is transmitted through the facet and becomes the output laser beam, this is another loss mechanism of the laser. The remaining portion of the photons is reflected by the facet, travels backwards inside the active layer, and induces more electrons and hole recombinations and more photon emissions. This process continues if there is enough electron and hole supply. The gain magnitude is proportional to the supply of electrons and holes or the injection current magnitude.

Although homojunction laser diodes are no longer produced, the homojunction structure is still adequate to be used to illustrate the working principle of laser diodes.

1.1.2   Gain-Guided Laser Diodes

Since the active layer structure shown in Fig. 1.1 is sandwiched inside another material, the laser field is confined in the y direction, vertical to the active layer. In the horizontal x direction, the laser field is confined by two mechanisms:

1. The gain profile. The width of the injection current is limited by the width of the top metal contact, as shown in Fig. 1.1. Outside the current flow area, there is no gain. The laser field will be absorbed by the active layer material. This mechanism provides a weak field confinement.
2. Gain-guided mechanism. Laser diodes have a unique property; the refractive index of the active layer is a complex function of the electron and hole densities inside the layer. As the current flows through the active layer, the index is increased by a small amount, ~ 0.01 or so depending on the current intensity and the material involved. This small index variation provides another weak field confinement. Therefore, the laser diode shown in Fig. 1.1 is called gain-guided laser diode. The field confinement of gain-guided laser diodes is too weak, the resulted beam profiles in the horizontal direction (x direction in Fig. 1.1) are not near Gaussian, which is undesirable. Gain-guided laser diodes can neither well confine the electrons and holes inside the active layer. Therefore, the electron and hole densities inside the active layer are low, resulting in a low recombination rate, low lasing efficiency, and high lasing threshold gain. Because of these shortcomings, gain-guided laser diodes are no longer produced; index-guided laser diodes are invented to improve the performance.

1.1.3   Index-Guided Laser Diodes

Double heterojunction laser diodes were later invented to solve the high threshold and low efficiency problems of gain-guided laser diodes. Figure 1.3a shows the schematic of the front view of a buried double heterojunction active layer. There are two more layers outside a homojunction to form a heterojunction. The material surrounding the homojunction has a refractive index lower than the index of the homojunction and can more tightly confine the electrons, holes, and the laser field in the horizontal and vertical directions. The better confinement raises the electron and hole densities inside the active layer, and thereby raises the lasing gain and efficiency. This type of laser diode is called the index-guided laser diode.

1.1.4   Quantum Well and Multi-quantum-Well Laser Diodes

Replacing the p-n homojunction shown in Fig. 1.3a by a quantum well, we obtain a buried heterostructure single quantum well laser diode. A quantum well is only around 10 nm thick, much thinner than the ~ 0.1 μm thickness of a homojunction,

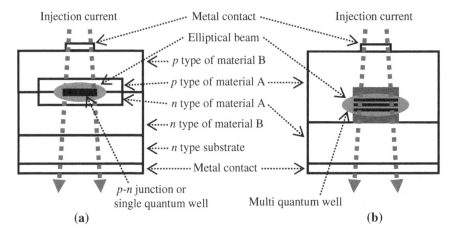

Fig. 1.3 Schematic of the front view of **a** a buried double heterojunction or a buried heterostructure single quantum well laser diode and **b** a buried homostructure multi-quantum-well laser diode

and is also sandwiched inside a different material. This feature of a quantum well leads to higher electron and hole densities than in a heterojunction. Thus, quantum well laser diodes have higher lasing efficiency and lower threshold gain.

Because the thickness of a quantum well is very thin, the net quantities of electrons and holes inside the quantum well are small, although the densities are high. This means the net output laser power from a quantum well can also be small, although the lasing efficiency is high. Multi-quantum-well laser diodes consisting of a number of parallel quantum wells are then developed to raise the net output power. Figure 1.3b shows the schematic of a buried homostructure multi-quantum-well laser diode.

The band structure of a multi-quantum-well laser diode is shown in Fig. 1.4; every quantum well is identical. As electrical current flows through the quantum wells, the electron density reduces significantly caused by both the recombination with holes and the material resistance. The electron densities must be very high in the first few quantum wells in order to have high enough densities in the last few

Fig. 1.4 Band structure of a
multi-quantum-well laser
diode

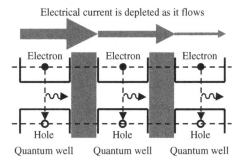

quantum wells for proper lasing. The excessive electrons in the first few quantum wells will cause high power consumption and heating. This shortcoming is improved in quantum cascade lasers to be discussed in the next section.

1.1.5 Intersubband and Interband Quantum Cascade Lasers

1.1.5.1 Intersubband Quantum Cascade Lasers

The same as a multi-quantum-well laser diode, a quantum cascade laser also consists of several quantum wells in parallel, as shown in Fig. 1.5. The differences are:

1. The quantum wells in a cascade laser are so thin that there are discrete quantum energy levels inside each conduction band. A photon is emitted when an electron transmits from a higher energy level to a lower energy level inside the same band. This process is called "intersubband."
2. Each quantum well is so designed that the energy levels inside the band step-down from one quantum well to the next in the direction of the electrical current flow, as shown in Fig. 1.5. The electrons that undergo an intersubband transition in one band can tunnel into the next quantum well, and undergo another intersubband transition and emit another photon.

These two differences bring quantum cascade lasers two advantages over multi-quantum-well laser diodes:

1. The energy levels are primarily determined by the thicknesses of the quantum wells, rather than by the materials used. Changing quantum well thickness to change the lasing wavelength over a large range is easier than changing materials to change wavelength. Quantum cascade lasers can emit in the middle IR range.
2. In intersubband quantum well lasers, electrons undergoing one transition and emitting photons in one quantum well can continue transition and emitting photons in the next quantum well, and so on. One electron can emit multiple photons. Electron density is reduced only by resistance as the current flows

Fig. 1.5 Band structure of an intersubband cascade quantum well laser

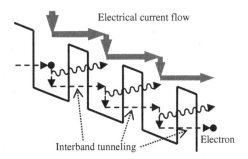

through the quantum wells. There is no need to keep very high electron density in the first few quantum wells. Therefore, quantum cascade lasers have lower threshold and higher efficiency than quantum well laser diodes.

Since there are no *p-n* junctions, quantum cascade lasers are not laser diodes, but still semiconductor lasers.

1.1.5.2 Interband Quantum Cascade Laser Diodes

Similar to intersubband quantum cascade lasers, an interband quantum cascade laser diode has a cascade band structure, the energy level steps-down from one quantum well to the next in the direction of the electrical current flow, as shown in Fig. 1.6. The difference is that every band is a quantum well with a *p-n* junction. Auger recombination occurs in the *p-n* junction. A portion of the energy emitted during the transition of an electron is converted into the emission of a photon; another portion of the energy is transferred to the second electron. The second electron tunnels into the next quantum well with lower energy level and Auger recombination occurs there again. This process continues, one electron can emit multiple photons, and the electron density is reduced only by resistance as the current flows through the quantum wells. Therefore, interband quantum cascade laser diodes also have low threshold, high efficiency, and can operate in the middle IR range.

1.1.6 DFB and DBR Laser Diodes

All these laser diodes described above have a large linewidth of ∼ 0.1 nm, when they are operating at single longitudinal mode. The corresponding coherent length is only of ∼ 10 mm. The coherent length can be one order smaller if the laser diode is operating at multilongitudinal modes. Such a short coherent length excludes these laser diodes from interference applications. Distributed Feedback (DFB) laser diodes and Distributed Bragg Reflector (DBR) laser diodes are invented to reduce the linewidth of laser diodes.

Fig. 1.6 Band structure of an interband cascade quantum well laser diode

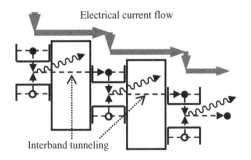

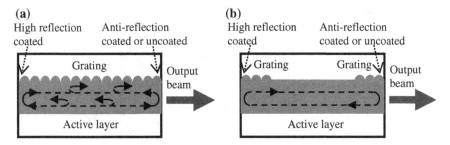

Fig. 1.7 Schematic of the side view of **a** a DFB laser diode and **b** a DBR laser diode

Figure 1.7a shows the schematic of the side view of a DFB laser diode. There is a periodic index structure (grating) built into the active layer. The grating is a narrow band reflector, and only the wavelength that equals the grating period can resonate and lase. The grating does not completely reflect the beam. One facet of the active layer still needs to be high reflection coated to provide total reflection. Another facet of the active layer can either be left uncoated or antireflection coated, depending on the strength of the grating reflection, to let the beam output. Because grating is distributed along the active layer, reflection takes place along the active layer as the beam is traveling inside the active layer, as shown in Fig. 1.7a. This mechanism can only be analyzed by coupled wave theory [1]. This is beyond the scope of this book. The wavelength-selective reflection of distributed grating can significantly reduce the laser linewidth. Different types of DFB laser diodes can have very different linewidths, ~ 1 MHz or $\sim 10^{-5}$ nm can be used as estimation.

Figure 1.7b shows the schematic of the side view of a DBR laser diode. There are two sections of periodic index structure (Bragg reflector) built into the active layer near the two facets. Similar to DFB laser diodes, the two Bragg reflectors provide a wavelength-selective reflection and thereby significantly reduce the laser linewidth. One facet of the active layer is high reflection coated and another facet of the active layer is either antireflection coated or uncoated to let the beam output. Different types of DBR laser diodes can have very different linewidths, typically a few times larger than the linewidth of DFB laser diodes because of the shorter grating.

1.1.7 Vertical Cavity Surface Emitting Laser Diodes (VCSEL)

VCSEL is a unique type of laser diode. As shown in Fig. 1.8, two Bragg reflectors sandwich an active layer and form the lasing cavity. A circular window is itched on one Bragg reflector to let the laser beam output. The output beam is thereby circular, which is the main advantage of VCSELs. The lasing cavity length equals the thickness of the active layer and is very short, the laser field does not have a lot

Fig. 1.8 Schematic of a
VCSEL

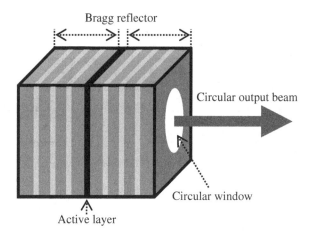

gain before outputting from the cavity, and the laser power can be only a few mW. This is the main disadvantage of VCSELs. The only way to increase the laser power is to increase the window size. However, large window size will result in multi-transverse modes in the output beam that takes away the only advantage of VCSELs.

1.1.8 Other Terminologies Often Used to Categorize Laser Diodes

Several different terminologies are often used to refer the same type of laser diode and may cause confusion. Here we briefly explain these terminologies.

1. Fabry-Perot laser diode. The two facets of the active layer in a laser diode form a resonant cavity for lasing. The two facets are parallel to each other and are similar to a Fabry-Perot echelon. All the laser diodes described above are Fabry-Perot laser diodes, except the DFB, DBR, and VCSEL laser diodes.
2. All the laser diodes described above, except the VCSEL laser diodes, emit beams from the edge of the active layer, and can be called edge emitting laser diodes.
3. Since laser power is generated by injecting electrons and holes into the active layer, all the laser diodes described above can be called injection current laser diodes.
4. Ridge waveguide laser diodes. Figure 1.9 shows the schematic of a multi-quantum-well laser diode with a ridge waveguide. The ridge waveguide provides a strong index guide, and can more tightly confine the laser field in the horizontal direction which results in lower threshold and a near-Gaussian beam profile in the horizontal direction. The ridge waveguide structure is widely used in laser diodes.

Fig. 1.9 Schematic of a ridge
waveguide laser diode

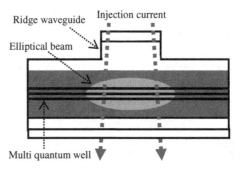

5. Wide stripe and broad area laser diodes. As explained earlier, the active layer
   must be kept thin in order to raise the electron and hole densities to lower the
   threshold gain and raise the lasing efficiency. Laser power can be increased by
   extending the active layer width. Wide stripe or broad area laser diodes refer to
   an active layer at least tens of microns wide. Such laser diodes usually emit
   multitransverse mode beams with over 100 mW power.

1.2 Gain

1.2.1 Lasing Threshold Condition

As the laser field travels back and forth inside the active layer, it gains strength from
stimulated emission, and loses strength due to material absorption and reflection
loss of the output facet. Once the gain is equal to or larger than the total loss, lasing
starts. The lasing threshold condition is given by Coldren and Corzine [2]

$$R_1 R_2 e^{2(g-\alpha_i)L} = 1 \tag{1.1}$$

where R_1 and R_2 are the reflectivity of the two facets, respectively, g and α_i are the
gain and loss per unit length along the active layer, respectively, L is the physical
length of the active layer, and the left-hand side of the equation is the sum of all the
gains and losses the laser field undergoes after the laser field makes one round trip
inside the cavity. The lasing threshold condition of Eq. (1.1) is a general threshold
condition and holds for all types of lasers. For DFB and DBR laser diodes, R_1 and
R_2 are complex numbers, which means they also introduce a phase change to the
laser field.

Table 1.1 Commonly used combinations of semiconductor materials and their gain profiles

Material combinations	Central wavelength (μm)	Spectral bandwidth (nm)
GaN/InGaN, GaN/AlGaN	0.3–0.5	~2
GaAs/GaAlAs	0.52–0.98	~20
InP/InGaAsP	0.9–1.65	~80
GaSb/GaInSb, InAs/GaSb/InAs/ AlSb	2–12	>100

The abbreviations in the tables are
GaN/InGaN Gallium nitride/Indium gallium nitride
GaN/AlGaN Gallium nitride/Aluminium gallium nitride
GaAs/GaAlAs Gallium arsenide/Aluminium Gallium arsenide
InP/InGaAsP Indium phosphide/Indium Gallium arsenide phosphide
GaSb/GaInSb gallium antimonide/gallium Indium antimonide
InAs/GaSb/InAs/AlSb Indium arsenide/gallium antimonide/Indium arsenide/Aluminum antimonide

1.2.2 Laser Diode Materials and Gain Profiles

Most laser diodes nowadays are heterostructural. A combination of two types of materials forms a heterostructure. Different combinations of two materials have different gain profiles.

A gain profile can be approximated by a Gaussian or Lorentz function and is specified by its central wavelength and spectral bandwidth. A laser diode will naturally lase at one or several wavelengths around the center of the gain profile. The spectral bandwidth provides the potential of tuning the lasing wavelength inside the bandwidth if there is a wavelength tuning mechanism applied to the laser diode. Below is a table listing commonly used material combinations and their gain profiles (Table 1.1).

1.3 Spectral Properties

There are many types of laser diodes with different materials and different active layer structures. Even for the same type of laser diodes, the manufacturing tolerances can cause these laser diodes to behave differently from each other. The situation for laser diodes is more complex than many other types of lasers.

1.3.1 Longitudinal Modes

Both the longitudinal and transverse modes are often abbreviated as "mode." But these two modes have completely different meanings. Longitudinal modes refer to those laser fields with wavelengths supported by a laser cavity, while transverse modes refer to those laser fields with spatial structures supported by the laser cavity.

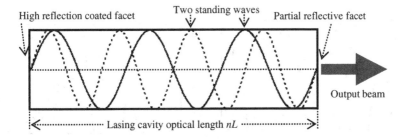

High reflection coated facet   Two standing waves   Partial reflective facet

Output beam

←·············· Lasing cavity optical length nL ··············→

Fig. 1.10 Illustration of two standing waves in a laser cavity

The optical length of a resonant cavity of a laser diode is given by nL, where n is the refractive index of the cavity and L is the physical length of the cavity. A cavity allows many standing optical waves or longitudinal modes to exist in it. Figure 1.10 shows two standing waves as examples. Any longitudinal mode must meet the condition

$$m\lambda/2 = nL \qquad (1.2)$$

where m is the mode order which is an integer, and λ is the lasing wavelength. Differentiating Eq. (1.2) and assuming $\Delta m = -1$, we obtain $\Delta\lambda$, the difference between two neighboring standing waves or the mode spacing

$$\Delta\lambda = -\frac{\Delta m}{m}\lambda = \frac{\lambda^2}{2nL} \qquad (1.3)$$

For laser diodes, n typically between 3 and 4, and L typically a fraction of mm, $2nL$ can be estimated to be 1 mm. We obtain $\Delta\lambda = 1$ nm from Eq. (1.3), assuming $\lambda = 1$ μm.

1.3.2 Mode Competition

A resonant cavity can support many longitudinal modes as shown in Fig. 1.11a. But only those modes inside the gain band are amplified by the active medium, as shown in Fig. 1.11b. The longitudinal mode closest to the gain peak receives the largest gain and is the strongest mode. The strongest mode consumes more electrons and holes than the other modes and leaves fewer for other modes to consume, since the numbers of electrons and holes are limited. Thus the strongest mode is in a more advantageous position. This advantage position makes the strongest mode even stronger and consumes even more carriers. This process continues till eventually all the other modes are extinguished. Such a phenomenon is called "mode competition." Mode competition enables most laser diodes operating at single longitudinal mode or single wavelength, which is good news for laser diode users. However, multilongitudinal modes operation is not rare for laser diodes.

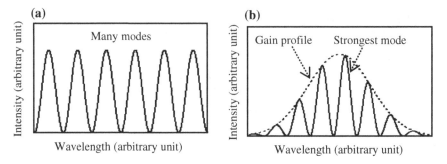

Fig. 1.11 **a** A laser cavity can support many longitudinal modes. **b** Only those longitudinal modes inside the gain profile are supported by the active medium

Particularly at low power level, the gain profile is relatively flat; the strongest mode does not have enough advantage over the other modes to extinguish them.

Figure 1.12 shows a typical mode structure change when the power of a laser diode is tuned up. At low power level, there are several modes. As the power is increased, the laser diode eventually operates at a single mode. The central wavelength change shown in Fig. 1.12 is a result of gain profile shift caused by temperature rise. This phenomenon will be explained later.

1.3.3 Mode Hopping

As we can see from Eq. (1.2) the wavelength of a longitudinal mode is proportional to the optical length of the cavity nL. The refractive indexes n of the active mediums is a function of the carrier (electrons and holes) density or inject current, the wavelength thereby can be changed by changing the injection current magnitude. A typical changing rate is $d\lambda/dI \approx 0.01$ nm/mA. Thermal expansion of the active layer will also change the wavelength. As the temperature of the active layer and/or the current injected into the active layer changes, both the gain profile and

Fig. 1.12 Typical mode structure change versus power change

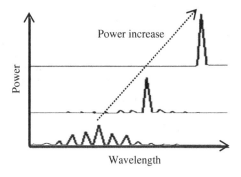

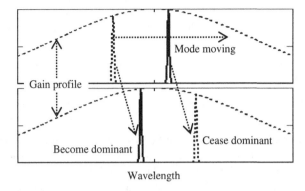

Wavelength

Fig. 1.13 Illustration of mode hopping. The dominant mode moves away from the gain peak and ceases dominant. A non-dominant mode moves closer to the gain peak and becomes dominant

the longitudinal modes shift and their relative positions change, as shown in Fig. 1.13. The strongest mode may gradually move away from the gain peak and another mode may gradually move toward the gain peak. After passing through a critical point, the other mode becomes dominant and extinguishes the previous strongest mode. This phenomenon is called "mode hopping." Sometimes, a laser diode is operated at such a state that two modes are equally close to the gain peak, a small perturbation in temperature and/or current can tip the balance, repeated mode hopping between these two modes will occur. Mode hopping changes the wavelength and laser power, even the beam spatial profile can be slightly affected. Continuous mode hopping is annoying and should be avoided by slightly adjusting the temperature or injection current of the laser diode.

1.3.4 Wavelength

Laser diode wavelength ranges from ultraviolet to infrared. The wide wavelength selection is one of the major advantages of laser diodes over other types of lasers. Because of the manufacturing tolerance, laser diodes usually have a wavelength tolerance of ±5 nm or so. Laser diode users should be fully aware of this. If they want a more accurate wavelength, they can ask laser diode vendors to select these laser diodes with the right wavelengths for them with a premium.

Some commonly used laser diode wavelengths are listed below:

375 nm	405 nm	445 nm	473 nm	485 nm	510 nm	635 nm
640 nm	657 nm	670 nm	760 nm	785 nm	808 nm	848 nm
980 nm	1,064 nm	1,310 nm	1,480 nm	1,512 nm	1,550 nm	1,625 nm
1,654 nm	1,877 nm	2,004 nm	2,330 nm	2,680 nm	3,030 nm	3,330 nm

Since laser diode technology is progressing fast, the wavelength selection range can be extended any time. Laser diode users should check for the latest development in this field.

1.3.5 Linewidth

There are three main contributors to the linewidth of a laser diode. They are spontaneous emission, amplitude to phase coupling described by the "linewidth enhancement factor," and the $1/f$ noise. The FWHM linewidth Δv of a single longitudinal mode laser diode is given by the Schawlow-Townes-Henry equation [3]

$$\Delta v = \frac{h v v_g^2 (\alpha_i + \alpha_m) \alpha_m n_{sp}}{8 \pi P_{\text{out}}} (1 + \alpha_H^2) \tag{1.4}$$

where h is the Planck constant, v is the frequency, v_g is the group velocity of the laser field inside the active layer, α_i is the active layer material per length loss, α_m is the cavity mirror loss, n_{sp} is the population inversion factor, P_{out} is the output power, and α_H is the "linewidth enhancement factor." The cavity mirror loss is given by $\alpha_m = -\ln (R_1 R_2)/(2L)$, where L is the laser cavity physical length, $R_1 \approx 0.3$ is the reflectivity of the uncoated facet of the active layer which the laser outputs and $R_2 \approx 1$ is the reflectivity of the high reflection coated facet. Equation (1.4) indicates that the linewidth of a laser diode is inversely proportional to the cavity length L and the output power. A typical value of laser diode linewidth is $\Delta \lambda \sim 10^{-1}$ nm. The coherent length ΔL_c of a laser diode beam is related to its linewidth by $\Delta L_c = \lambda^2/\Delta \lambda \sim 10$ mm, if we assume the wavelength to be $\lambda \approx 1$ µm. The short coherent length of laser diode beams limits their interference-related applications.

As described in Sect. 1.1.6, DFB and DBR laser diodes have narrower linewidth. Another way of significantly reducing the linewidth is to construct an external cavity for a laser diode. An external cavity can increase the cavity length by tens to hundreds of times and dramatically reduce the linewidth according to Eq. (1.4). We will discuss the external cavity laser diode in Sect. 3.7.

1.4 Power

1.4.1 Continuous Wave Operation

Laser diodes can output continuous wave (CW) laser power ranging from a few mW for a single transverse mode laser diode to several kW for a laser diode stack. A typical laser power versus electrical inject current curve looks like as shown in Fig. 1.14a, this curve is called $L \sim I$ curve. The slope efficiency is defined as

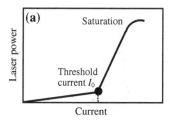 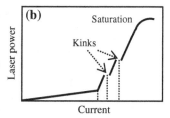

Fig. 1.14 **a** A typical $L \sim I$ curve. **b** A $L \sim I$ curve with two kinks

$$\eta = \frac{P}{I - I_0} \qquad (1.5)$$

where P is the laser power, I and I_0 are the operation and threshold currents, respectively. When the current is below the threshold, the laser diode only has spontaneous emission; that is, a weak and widespread light with the same wavelength as the laser light. The spontaneous emission intensity slowly increases as the current increases. At the threshold, the lasing gain equals the sum of all the losses. When the injection current is above the threshold, the laser diode starts lasing, the laser power increases fast as the current is increased. As the current continues increases, the slope of the curve will gradually reduce; the laser power reaches its saturation level, as shown in Fig. 1.14a.

Laser diode users should slowly increase the current till the laser power reaches the level specified by the datasheet to avoid overdriving the laser diode. A fraction of second of overdriving can partially or totally blow out the laser diode. The threshold current level is from tens to hundreds of mA for different types of laser diodes. Again, different laser diodes of the same type can have tens of percents difference in threshold current and tens of percents difference in slope efficiency.

Sometimes, an $L \sim I$ curve has kinks, as shown in Fig. 1.14b. Kinks are a phenomenon of longitudinal mode hopping, because changing current will change the optical length of the laser cavity and the temperature, and cause mode hopping or kink. For CW applications, the laser current should be adjusted away from the kink point. For modulated applications, the current swipes over a range that may cover the kink points, then kinks may affect the applications. Some laser diodes are guaranteed kink free within their specified operation ranges.

1.4.2 Modulated or Pulsed

In some applications, such as fiber optic communications, the light of a laser diode is modulated to carry signals. Modulation can be performed by modulating the injection current. When the injection current is changed, the laser field needs to take several round trips inside the laser cavity to establish the new operation state. The time needed to make these several round trips sets the theoretical maximum

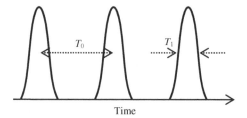

Fig. 1.15 Three pulses with period T_0 and width T_1

modulation rate for a laser diode. For a laser cavity with an optical length of $nL = 1$ mm, one round trip inside the cavity takes $2nL/c = 2 \times 1$ mm/ $(3 \times 10^{11}$ mm s$^{-1}) \approx 7 \times 10^{-12}$ s, where c is the velocity of light in a vacuum. Several round trips take $\sim 10^{-11}$ s or $\sim$ GHz in term of frequency. This number agrees with the practically achieved maximum modulation rate of laser diodes. Such a high frequency modulation is mainly used in fiber optic telecommunications. In most other applications, megahertz modulation speed will be fast enough and is quite comfortable for laser diodes.

In some other applications, such as ranging, a laser diode is pulsed to provide peak power higher than the continuous power. The smallest pulse width and the highest repetition rate are limited in a way similar to the limitation on the maximum modulation rate. There are many types of laser diodes specially designed for pulsed operation, the typical pulse width and duty cycle is 100 ns and 0.1 %, respectively. A pulsed electrical power source is required to pulse a laser diode. Pulsed laser diode modules have special electronics built in the modules; the electronics can take in CW electrical power and output electrical pulse.

Figure 1.15 shows three pulses with period T_0 and width T_1. The pulse width is usually defined as the FWHM. The energy per pulse E is often specified for a pulsed laser diode. We can calculate the peak and average power by $P_{\text{Peak}} = E/T_1$ and $P_{\text{Average}} = E/T_0 = E \times$ Frequency.

1.5 Temperature Effect

Laser diode operation states are affected by temperature change. High operation temperature will increase the threshold current by $I_{th} \sim \exp(T/T_0)$, where $T_0 \approx 50$ °K ~ 200 °K is the characteristic temperature, and reduce the electrical to optical conversion efficiency at a rate of a few percents per degree °C, as shown in Fig. 1.16. Laser diode life expectancy LE and the temperature increment ΔT is related by $LE \sim \exp(-T_0 \Delta T/T^2)$, where $T_0 \approx 2300$ °K ~ 8000 °K is a characteristic temperature. Temperature rising will significantly reduce the life expectance. Thermal expansion of the cavity can increase the wavelength, with a typical change rate of $d\lambda/dT \approx 0.2$ nm/°C in visible range and $d\lambda/dT \approx 0.3$ nm/°C in infrared. Temperature change will also change the value of the current at which a kink occurs. Low-power

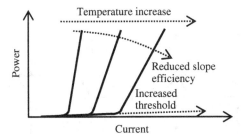

Fig. 1.16 Increasing temperature will increase the threshold current and reduce the efficiency of the laser diode

laser diode packages have a metal base of about 100 mm³ volume in it. This metal base is the mounting support as well as the heat sinker, and should be in thermal contact with a larger heat sinker when the laser diode is operated. A laser diode active layer has a tiny volume of 100 μm³ or so and the electrical current and laser power densities are very high inside such a small volume. The temperature of the active layer can be over 100 °C and can be measured by measuring its IR spectrum, which is not a very easy measurement. Laser diode temperature is often indirectly monitored and controlled by monitoring and controlling its heat sinker temperature. The therrmal impedance between laser diode active layer and the heat sinker is ~ 10 °C/W. As the temperature of a laser diode rises, the maximum allowed current and laser power drops. A driving current safe at lower temperature may be too large at higher temperature.

High laser power density and high temperature are the main failure causes of laser diodes. If proper protection procedures are taken, the laser diode lifetime is about 10,000 h.

1.6 Electrical Properties

1.6.1 Internal Circuit for DC Operation

A laser diode can be forward driven by a DC power source of about 2 V or higher such as a battery. Laser diode datasheet should specify the power source voltage. The electrical to laser energy conversion efficiency of laser diodes is defined by Eq. (1.5) and is in the range of 0.5 mW/mA. The internal circuit of laser diodes can be slightly different for different types of laser diodes. Figure 1.17 shows three typical internal circuits in a laser diode can. The photodiode in the internal circuit is physically mounted behind the back facet of a laser diode active layer to pick up the tiny laser power leaking through the high reflection coated back facet. The photodiode output current is proportional to the laser power and is used as a feedback in a circuit loop to stabilize the laser power. It is noted that a laser diode module cannot be simultaneously operated at power stabilized mode and current stabilized mode without temperature control. Because at a constant current level, temperature

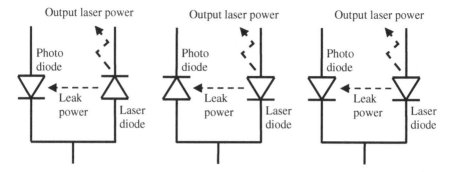

Fig. 1.17 Three types of internal circuits in a laser diode can

change will change the laser power, or at a constant power level, temperature change requires different current. Power stabilization is one of the basic functions of laser diode modules. Laser diodes users should check the datasheet to find how the internal circuit is connected to the pins of the package.

The circuits shown in Fig. 1.17 are only adequate for DC operation of laser diodes.

1.6.2 Series Resistance

The series resistance R describes the DC behavior of a laser diode and can be found by measuring the voltage v applied to and the current i passing through the laser diode, as shown in Fig. 1.18a, we have $R = dv/di$. A typical $v \sim i$ curve and $R \sim i$ curve is shown in Fig. 1.18b. It is noted that R is not a constant. Below the threshold, the value of R drops fast as the voltage and current increase. Above the threshold, the value of R slightly drops as the voltage and current increase, because

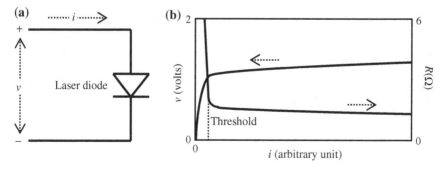

Fig. 1.18 **a** A laser diode is forward biased. **b** A typical $v \sim i$ curve and $R \sim i$ curve for a laser diode

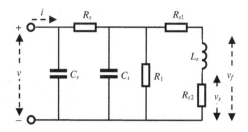

Fig. 1.19 One equivalent circuit for a laser diode

the optical and electrical processes inside a laser diode are related. For example, there is an equivalent resistor R_1 in the circuit model of the electrical properties of a laser diode, as shown in Fig. 1.19 in the following section. Reference [4] shows that $R_1 \sim 1/N_0$, where N_0 is the carrier density inside the laser diode. Below the threshold, N_0 increases fast as the current increases, so R_1 drops fast. Above the threshold, as the current increases, the increased N_0 is converted into laser power, the net value of N_0 is only slight increased, and R_1 only drops slightly.

1.6.3 Intrinsic Circuit for Modulation

When modulated, the complex optical process inside a laser diode will affect the way the laser diode responds to the driven circuit. The parasitic circuit will also affect the electrical characteristics of a laser diode when the modulation frequency is high. Therefore, when modulated, the electrical property of a laser diode is much more complex than a series resistor and can be modeled by an equivalent circuit. Many papers on this subject have been published. These papers all use the rate equations to address various aspects of various laser diodes and obtain different equivalent circuits.

Figure 1.19 shows an equivalent RLC circuit of a laser diode obtained by using small signal approximations [4]. In this equivalent circuit, v and i are the external applied voltage and current, respectively. C_s and R_s are the contact capacitance and resistance between the contact and the active layer, respectively. R_{s1}, R_{s2}, and L_s are two equivalent resistors and one equivalent inductor, respectively; R_{s1}, R_{s2}, and L_s are functions of optical parameters of the laser diode and model the spontaneous emission effect. R_1 and C_t are the equivalent resistor and capacitor, respectively. R_1 and C_t are also functions of optical parameters of the laser diode and model the diffusion effect of the carriers. The voltage v_s across R_{s2} shown in Fig. 1.19 can be used as a measure of the optical intensity. The voltage v_f across R_{s2} and L_s is a scaled analog of the small signal frequency modulation.

For large signal modulation, the approximations taken for small signal analysis are no longer effective; the mathematical process of solving the rate equations becomes more complex. There are papers published on this subject, for example Ref. [5].

1.7 Main Failure Mechanism and How to Protect Laser Diodes

Laser diodes are extremely sensitive to electrostatic discharge, excessive current levels, and current spikes or transients. The workstation and the person handling laser diodes should be properly grounded. Otherwise, a laser diode can be damaged without any noticeable sign. Being vulnerable to surges in the injection current, laser diodes should be driven by a current source or batteries through a protective circuit or by a special power supply. Once a laser diode is mounted in a module, it is safe to handle the module by ungrounded hands or tools. The module housing will shield the laser diode inside from electrical static discharge, and is safe to operate the module by a conventional power source. The electronics inside the module will protect the laser diodes from power surge.

The laser power density at the emission facet can be very high, of the order of ~ 10 mW/μm^2 or ~ 10 GW/mm^2. Any tiny defects or pollution on the facet can cause power absorption, heating, and oxidizing the facet, and eventually damage the facet. Therefore, do not try to use conventional method to clean the facet, this may scratch the facet or pollute the facet even more.

When directly handling laser diodes, electrostatic discharge is the number one cause of damage, followed by accidental disconnection of power, and then overheating. When handling laser diode modules, the main cause of damage is overdriven. Symptoms of damage include reduced output power, increased threshold current, increased beam divergence, difficulty in focusing to previously attained spot sizes, and ultimately failure to lase. A damaged laser diode often behaves like an LED device, only outputs weak light, not intensive beam.

Reference [6] provides a detailed discussion on protecting laser diodes.

1.8 Laser Diode Mechanical Properties, Packages, and Modules

1.8.1 Mechanical Tolerance

Laser diodes have manufacturing tolerances much larger than the tolerances of other types of lasers. In previous sections, we have already mentioned that laser diodes have a wavelength tolerance of several nanometers, a threshold current tolerance of tens of percents, and an $L \sim I$ curve slope tolerance of tens of percents. Besides these tolerances, laser diode chips also have a mechanical mounting tolerance of about ± 0.1 mm in all the three directions inside the laser cap, and a pointing tolerance of about $\pm 0.2°$. Figure 1.20 depicts a chip mounted with transverse position and pointing tolerance.

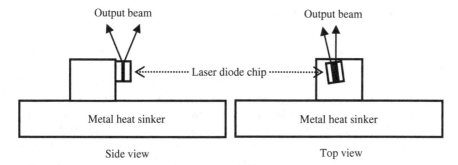

Fig. 1.20 Schematic of a laser diode chip mounted on a metal heat sinker, not drawn to the *right* proportion for illustration purpose. The mounting tolerance in the *top* view is exaggerated for clarity

1.8.2 Laser Diode Packages

Low-power laser diodes have two standard packages; 9 or 5.6 mm cap, the two packages have similar shapes and proportions and only the sizes are different, as shown Fig. 1.21a. A laser diode chip is mounted inside the cap. The cap is filled with inert gas and sealed. The inertial gas will significantly slow down the oxidization of the laser diode facet caused by the high laser power density and high temperature. A sub mm thick glass window on top of the cap lets the laser beam output. The cap size has nothing to do with the power of the laser diode chip inside. The metal base of laser diode caps serves as a mounting base and as a heat sinker.

There are no standard packages for high-power laser diodes or laser diode stacks. So, do not be surprised to see packages with various shapes, structures, and sizes. Figure 1.21b shows some of the packages. High-power laser diodes sometimes require additional cooling in addition to the heat sinkers coming with the packages.

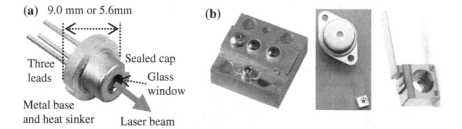

Fig. 1.21 a A standard low-power laser diode package. **b** Some types of high-power laser packages

1.8.3 Laser Diode Modules

Since the beams emitted by laser diodes have high divergence and are difficult to
manipulate, and laser diodes are vulnerable to static discharge, laser diodes are
often sold as modules with collimated beams and protection housings. Laser diode
users only need to handle the modules instead of directly handling laser diodes.

A typical laser diode module contains at least four basic components: a laser
diode, a collimating lens(es), a circuit board, and a housing to hold all the components together. The housing is usually a sealed metal tube with two or more wires
for connecting the laser diode and circuit board inside the housing to an electrical
power source, as shown in Fig. 1.22. The lens collimates the laser diode beam so
that laser diode users can have a collimated beam to work with. The collimating
lens may be movable along the optical axis to provide adjustable focusing capa-
bility. The circuit board has at least a laser power stabilizing function utilizing the
output of the photodiode mounted behind the laser diode, some circuit boards have
injection current stabilizing function and/or temperature stabilizing functions, and
some boards may have modulation or pulsation or even programming functions.
The circuit board shields the laser diode from electrical surge from the power
supply. The module may have an electrical fan attached to it to provide air cooling.
The metal houses of laser modules are electrically grounded to protect the laser
diode inside from static discharge.

Laser diode vendors offer a wide selection of laser diodes, collimating lenses,
and some selections of circuit boards for the users to choose from. Sometimes, they
can assemble the modules to best fit the user's needs. Buying a laser diode module
may cost twice more than buying these three components separately and assembling
them by the users themselves, but can save the user a lot of time and effort.

When assembling a laser diode module, the laser diode chip needs to be positioned at the focal point of the collimating lens. Even we can well position the laser
diode cap relative to the collimating lens; the laser diode chip inside the cap may
not necessarily be positioned at the focal point of the lens, because of all these
mechanical tolerances. Some type of position adjustment mechanisms must be
applied to the laser diode cap. Adjusting the axial position of a laser diode cap is

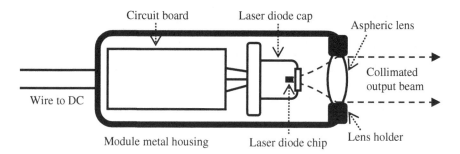

Fig. 1.22 Schematic of a most basic laser diode module

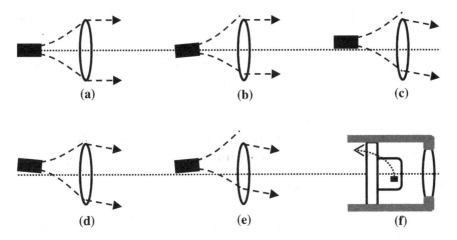

Fig. 1.23 **a** An ideal collimation situation. The laser diode chip is positioned at the focal point of the lens and has no boresight error. **b** A most common situation. The laser diode chip is positioned at the focal point of the lens, but has boresight error and the collimated beam has no boresight error. **c–e** The laser diode chip is not at the focal point of the lens, the collimated beam has boresight error. **f** A laser diode cap is well positioned inside a housing, but the chip is not at the focal point of the lens. The chip can be moved to the focal point by rotating the cap

relatively simple. Adjusting the transverse position of a laser diode cap is more complex, we explain the details in Fig. 1.23. The drawings in Fig. 1.23 are not to the right proportion for illustration purpose.

Figure 1.23a shows the ideal situation, with the laser diode chip at the focal point of the lens and with no boresight error. Figure 1.23b shows a less ideal situation, the laser diode chip is at the focal point of the lens, but has a boresight error. It is noted the collimated beam does not have a boresight error, but has a small transverse displacement and is more truncated at one side. If the laser diode chip has 0.2° boresight error and the collimating lens has a focal length of 5 mm, the transverse displacement will be 0.2° × 5 mm = 0.017 mm, which is trivial. The situation shown in Fig. 1.23b is the most common and the collimated beam is still of good quality. The laser diode chip shown in Fig. 1.23c–e has transverse position errors that will result in boresight errors in the collimated beam. If the transverse potion error is 0.1 mm and the lens focal length is 5 mm, the boresight error of the collimated beam is 0.1 mm/5 mm = 0.02 Radian, which may be an issue in some applications. The boresight error of the chip shown in Fig. 1.23c–e will affect the truncation of the beam, but will not affect the boresight error of the collimated beam, and is of less importance. Figure 1.23f depicts a real situation, where the laser diode cap is well positioned inside a module housing, but the laser diode chip is not at the focal point of the lens because of the manufacturing tolerance. The laser diode chip can be moved to the focal point of the lens by rotating the cap as shown in Fig. 1.23f, the final situation is similar to that shown in Fig. 1.23b. Rotating laser diode cap in any direction can be realized by using three set screws to push the cap, or by using some type of tool.

As the laser diode cap position is being adjusted, the far field beam pattern and beam position must be monitored. The alignment is done when the far field beam pattern is clean and symmetric, as shown in Fig. 3.6a.

1.9 Vendors and Distributors of Laser Diodes, Laser Diode Modules and Laser Diode Optics

1.9.1 Laser Diode Vendors

Fabrication of laser diode chips require expensive equipment and tight process control. Laser diodes produced by brand name companies are of much higher quality, particularly have much longer lifetime, but can cost twice more than those laser diodes produced by unknown companies. In most applications, brand name laser diodes are worth the price. It is not a pleasant experience to stop the whole operation process and have service personnel coming to fix expensive equipment only because the $100-worth laser diode in the equipment is blown out. Below is a list of brand name laser diode manufacturers:

*Applied Optronics   Coherent   Dilas Diodenlaser   Frankfurt Laser Company
Hamamatsu   Hitachi   JDS Uniphase   Jenoptik   Mitsubishi   Opnext
Opto Power   OSRAM   Rohm   Lasertechnik   Sony   Toshiba*

It is noted that the list is subjective and likely incomplete. There are likely many other good quality laser diode manufacturers.

Nowadays, low-power laser diodes are produced in millions, the unit price can be as low as a few dollars. Laser diode manufacturers usually no longer retail sell their laser diodes. Instead, they whole sell their laser diodes to distributors. On the other hand, one distributor usually retail sells laser diodes from many manufacturers. If you want to buy low-power laser diodes, you should first try to contact distributors. Below is an incomplete list of distributors:

*Blue Sky Research   Digi-Key   Edmund Optics   Lasermate   Lasertel
Power Technology   Thorlabs*

High-power laser diodes are used in more special applications by much smaller quantities. The unit price can be high and the manufacturers usually do retail sales. Distributors also sell many high-power laser diodes.

1.9.2 Laser Diode Module Vendors

The task of assembling laser diode modules is easier than fabricating laser diode chips. Therefore, the brand reputation is not as important for laser diode module manufacturers. But the brands of laser diodes and lenses used in the modules are still important. Laser diode module manufacturers only sell their own modules. Distributors sell some commonly used laser diode modules from a few manufacturers. Manufacturers usually can provide more insightful technical consulting to help the user to choose the right modules and provide better after sale technical support for their products. Below is an incomplete list of laser diode module manufacturers:

Blue Sky Research   Coherent   CVI Melles Griot   Lasermate   Lasertel
Micro Laser System   Point Source   Power Technology

1.9.3 Laser Diode Optics Vendors

The most widely used lenses for collimating or focusing laser diode beams are produced by LightPath Inc. These are molded glass aspheric lenses specially designed and fabricated for collimating laser diode beams. This lens series was produced by a company named Geltech Inc., which was acquired later by LightPath. LightPath still uses Geltech lens to name this lens series. This series lens is of high quality and expensive, and costs from $50 to $100 per lens.

There are other companies in Europe, Japan, Taiwan, etc., which make laser diode collimating lenses.

A few distributors resell many of these lenses. It appears to be more convenient to buy lenses from these distributors. Below is an incomplete list of distributors:

Edmund Optics   Optima Precision   Thorlabs

There are many different aspheric lenses in the market. Most of them are designed for other applications and will not well collimate laser diode beams. Users should be fully aware of this fact and only use those lenses specially designed for collimating or focusing laser diode beams.

References

1. Morthier, G., Vankwikelberge, P.: Chapter 3 Coupled-Mode Theory of DFB Laser Diodes, Handbook of Distributed Feedback Laser Diodes. Artech House Publishers, Boston (2013)
2. Coldren, L.A., Corzine, S.W.: Diode Lasers and Photonic Integrated Circuits, Equation 2.20, 38. Wiley, New York (1995)

3. Coldren, L.A., Corzine, S.W.: Diode Lasers and Photonic Integrated Circuits, Equation 5.151, 241. Wiley, New York (1995)
4. Tucker, R.S., Pope, D.J.: Circuit modeling of the effect of diffusion on damping in a narrow-stripe semiconductor laser. IEEE J. Quantum Electron. **QE-19**, 1179–1183 (1983)
5. Tucker, R.S.: Large-signal circuit model for simulation of injection-laser modulation dynamics. IEE Proc. I Solid State Electron Dev. **128**, 180–184 (1981)
6. Application Note 3: Protect your laser diode, ILX Lightwave. http://assets.newport.com/webDocumentsEN/images/AN03_Protecting_Laser_Diode_IX.PDF

Chapter 2
Laser Diode Beam Basics

Abstract The basic properties of single transverse mode and multi-transverse mode laser diode beams are reviewed. The characteristics of a laser diode beam propagating through optical elements is analyzed using three commonly used math tools: analytical tool thin lens equation and ABCD matrix, numerical calculation, and software tool Zemax. The emphasis is on using thin lens equation and numerical calculation to study the collimation and focusing characteristics of single transverse mode laser diode beams.

Keywords Astigmatism · Beam · Beam waist · Collimation · Divergence · Elliptical · Fast axis · Focal length · Focus · Gaussian · Image distance · Lens · M^2 factor · Object distance · Propagation · Rayleigh range · Raytracing · Slow axis · Spot size · Transverse mode

Single transverse mode laser diodes are most widely used. Their beams are elliptical, astigmatic, and have large divergence. These characteristics make laser diode beams difficult to handle. In this chapter we discuss in detail the basics of laser diode beams mainly using a simple paraxial Gaussian model. This model is accurate enough for most applications.

Multi-transverse laser diode beams are not typical laser beams and are also discussed in this chapter.

2.1 Single Transverse Mode Laser Diode Beams

2.1.1 Elliptical Beams

When a laser diode is operated, a portion of the laser field will transmit through one facet of the active layer and becomes the emitted laser beam. Because the active layer of a laser diode has a rectangular shaped cross section and a portion of the laser field will leak out from the active layer due to the limited confinement, the beam at the emission facet is a little larger than the cross section of the active layer

© The Author(s) 2015
H. Sun, *A Practical Guide to Handling Laser Diode Beams*,
SpringerBriefs in Physics, DOI 10.1007/978-94-017-9783-2_2

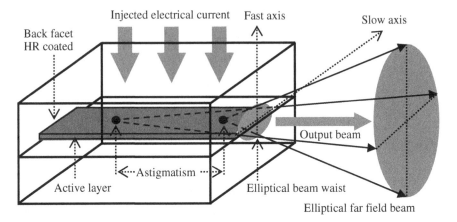

Fig. 2.1 A laser diode has a thin active layer. The emitted laser beam shown is *elliptical*, highly divergent, and astigmatic. The astigmatism magnitude is much exaggerated for clarity

and has an elliptical shape, as shown in Fig. 2.1. The beam size at the emission facet is about one micron in the direction vertical to the active layer and a few microns in the direction horizontal to the active layer. The beam elliptical ratio is typically from 1:2 to 1:4. The beam far field divergence is also different in the vertical and horizontal directions with a typical ratio of 2:1–4:1. Because the beam divergence is larger in the vertical direction, this direction is often called the "fast axis" direction. Then, the horizontal direction is called "slow axis" direction, as shown in Fig. 2.1.

As the beam propagates, the beam size in the fast axis direction will surpass the beam size in the slow axis direction, because the beam divergence is larger in the fast axis direction. The beam shape will become vertically elliptical, as shown in Fig. 2.1. This phenomenon is unique to laser diode beams. An elliptical shape beam is one of the undesired characteristics of laser diodes.

2.1.2 Large Divergences

The divergence of single transverse (TE) mode laser diode beams can vary significantly from different types of laser diodes and can even vary from diode to diode of the same type. The typical full width half magnitude (FWHM) divergent angle is about 15°–40° and 6°–12° in the fast and slow axis directions, respectively. In terms of $1/e^2$ intensity divergence, this is about 26°–68° and 10°–20° in the fast and slow axis directions, respectively. The laser diode industry traditionally uses the FWHM divergent angle to specify the beam divergence, because the FWHM number is more consistent; while in the optical community, the $1/e^2$ intensity divergence is often used. The latter is about 1.7 times larger than the former.

Because of the very large divergence in the fast axis direction, the lens used to collimate or focus a laser diode beam must have at least one aspheric surface to correct the spherical aberration and a numerical aperture (NA) of at least 0.3 to avoid severe beam truncation, although a lens with an NA of 0.6 will still truncate some beams. Most aspheric lenses specially designed and fabricated for collimating laser diode beams available in the market have an NA ranging from 0.3 to 0.6. Truncation of a beam will create side lobes, cause focal shift to the beam, and increase the divergence of the beam. The large divergent beam is another undesired characteristic of laser diodes.

2.1.3 Quasi-Gaussian Intensity Profiles

The spatial shape o of a laser diode beam is determined by the structure of the active layer. As described in Chap. 1, the active layer is one rectangular shaped waveguide or several rectangular shaped waveguides in parallel. TE modes of such active layers are not exactly Gaussian modes. The gain inside the active layer and the loss outside the active layer will also affect the mode shapes. There are many different active layer structures. The TE modes from these active layers are slightly different. There is no single mathematical model that can accurately describe all these modes and no commonly accepted relationship linking laser diode types to the shapes of their TE modes. Only individual case studies have been reported [1–3]. Based on this author's experience, in most practical applications the differences among most single TE modes of various types of laser diodes are insignificant and all these modes can be described with negligible error by a Gaussian model, since the Gaussian model is the simplest and most widely used model.

If we need be more specific, most single TE mode laser diode beams have slightly narrower central lobe and slightly longer tails compared with Gaussian mode.

2.1.4 Astigmatism

Laser diode beams are astigmatic; this is a consequence of the rectangular shaped active layer and the varying gain profile across the active layer in the slow axis direction. As shown in Fig. 2.1, the waist of a laser beam in the fast axis direction is located near the facet of the active layer, while the beam waist in the slow axis direction is located somewhere behind, that is, the astigmatism. The astigmatism depicted in Fig. 2.1 is much exaggerated for clarity. Similar to other laser diode parameters, astigmatism magnitude varies from different types of laser diodes and from diode to diode of the same type. For single TE mode laser diodes, the astigmatism is usually from 3 to 10 μm. For multi-TE mode laser diodes, the astigmatism is usually from 10 to 50 μm. From the application point of view, there is no need to study the origin of the astigmatism. We are more interested in measuring and correcting the astigmatism.

An astigmatic beam is another undesired characteristic of single TE mode laser diodes.

2.1.5 Polarization

Laser diode beams are linear polarized. The polarization ratio is high from about 50:1 to about 100:1 for single TE mode laser diode, and around 30:1 for wide stripe multi-TE modes laser diodes. The polarization is in the slow axis direction. The high polarization ratio of laser diode beams can be either an advantage or a disadvantage, depending on the type of applications. As a comparison, most He–Ne laser beams are randomly polarized.

2.2 Multi-transverse Mode Laser Diode Beams

2.2.1 Wide Stripe Laser Diode Beams

More carriers and photons are needed to increase the laser power. This can be achieved by increasing the volume of the active layer. However, as discussed in Chap. 1, high lasing efficiency requires high carrier density inside the active layer. This means the active layer thickness cannot be increased. Then, the straightforward way to increase the laser power is to increase the active layer width. For laser diodes with power higher than 100 mW or so (depending on the laser diode type and wavelength), the active layer widths are tens of microns or even up to a few hundred microns. Such laser diodes are often called wide stripe laser diodes or broad area laser diodes. The beam emitted from a wide stripe active layer contains multiple TE modes as depicted in Fig. 2.2. Every TE mode is a quasi-Gaussian mode. All these modes combine to form a multi-TE mode beam. As the beam

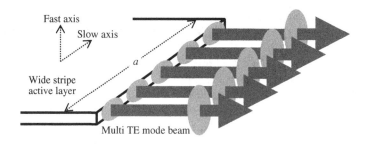

Fig. 2.2 The beam of a wide stripe laser diode contains multiple TE modes

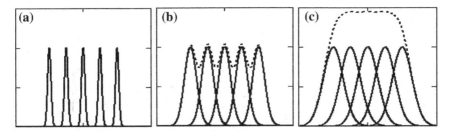

Fig. 2.3 The *solid curves* are for the spatial intensity distributions of five TE modes at three propagations distances. **a** At or near the diode facet. **b** At ten microns or so from the laser diode facet. **c** At tens of microns or beyond. The *dashed curves* are the spatial intensity distribution of five modes combined. The horizontal axis is spatial distance in the slow axis direction with a scale of tens of microns. The vertical axis is intensity with arbitrary unit

propagates, every mode increases its size and gradually merges with other modes to form a light line in the slow axis direction, as shown in Fig. 2.2. As the beam further propagates, the beam shape becomes rectangular, because all the modes have larger divergence in the fast axis direction.

Figure 2.3 shows the spatial intensity distribution of five TE modes at three different propagation distances. Figure 2.3a shows the intensity distribution of the five modes at or near the laser diode facet. As the beam propagates, the sizes of the five modes increase, the modes gradually merge together as shown in Fig. 2.3b, c by the thin curves. The intensity distributions of the five modes combined are shown in Fig. 2.3b, c by the dashed curves. When we scan such a multi-TE mode beam, the scan head is usually at least several millimeters away from the laser diode, the scan result will be something similar to that shown by the dashed curve in Fig. 2.3c. However, such a beam is not a true flat top beam. When the beam is focused, the intensity profile of the focused spot will be as shown in Fig. 2.3a if the focusing lens is of high quality, or like that shown by the dashed curve in Fig. 2.3b, where if the focusing lens has large aberration it will increase the size of the focused modes.

The beams of wide stripe laser diodes are not like the laser beams we have often seen, but are somehow like the lights from a light bulb. These beams cannot be well collimated or focused to small spots. We will discuss this in detail in Sect. 3.8.

2.2.2 Laser Diode Stack Beams

Several wide stripe active layers can be stacked up to further increase the laser power. Such a laser is called laser diode stack. There are many different combinations of active layer widths and stack layers. Figure 2.4 shows the schematic of a four-layer laser diode stack. There are many TE modes in the beam. The power of a laser diode stack can be up to thousands of watts. A laser diode stack can be treated

Fig. 2.4 Schematic of a laser
diode stack

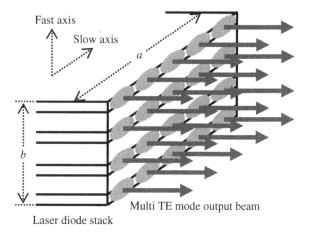

as a rectangular shaped light source of size $a \times b$ as shown in Fig. 2.4. The beams of
laser diode stacks are not like the laser beams we often see, but are rather like the
lights from a flashlight. These beams cannot be well collimated or focused to small
spots. We will discuss this in detail in Sect. 3.8.

2.3 Laser Diode Beam Propagation

2.3.1 Basic Mode Paraxial Gaussian Beams

Most laser beams have a circular shaped cross section with a Gaussian intensity
profile. Such beams are basic TE mode Gaussian beams. The characteristics of a
Gaussian beam can be described by a set of three equations [4]

$$w(z) = w_0 \left[1 + \left(\frac{M^2 \lambda z}{\pi w_0{}^2} \right)^2 \right]^{1/2} \tag{2.1}$$

$$R(z) = z \left[1 + \left(\frac{\pi w_0{}^2}{M^2 \lambda z} \right)^2 \right] \tag{2.2}$$

$$I(r, z) = I_0(z) e^{-2r^2/w(z)^2} \tag{2.3}$$

where $w(z)$ is the $1/e^2$ intensity radius of the beam at z, z is the axial distance from
the waist of the laser beam, w_0 is the $1/e^2$ intensity radius of the beam waist, M^2 is
the M square factor, λ is the wavelength, $R(z)$ is the beam wavefront curvature
radius at z, $I(r, z)$ is the beam intensity radial distribution in a cross section plane at
z, r is the radial coordinate in a cross section plane at z, and $I_0(z)$ is the beam peak

intensity in a cross section plane at z. The M square factor $M^2 \geq 1$ describes the deviation of the beam from a perfect Gaussian beam. For a perfect laser beam $M^2 = 1$. We will discuss the M^2 factor in detail in Sect. 2.3.2.

For a laser beam the Rayleigh range z_R is defined as that at $z = z_R$ the beam radius is $w(z_R) = \sqrt{2}\, w_0$. From Eq. (2.1) we can see that

$$z_R = \frac{\pi w_0^2}{M^2 \lambda} \tag{2.4}$$

z_R is proportional to w_0^2. From Eq. (2.1) we can also see that at far field, z is large, term $M^2 \lambda z / \pi w_0^2 = z/z_R \gg 1$, the $1/e^2$ intensity far field half divergent angle θ of the beam can be obtained by

$$\begin{aligned}
\theta &= \frac{w(z)}{z} \\
&= \frac{M^2 \lambda}{\pi w_0} \\
&= \frac{w_0}{z_R}
\end{aligned} \tag{2.5}$$

θ is inversely proportional to the beam waist w_0. Figure 2.5 plots Eq. (2.1) for two Gaussian laser beams with $w_0 = 1$ mm and $w_0 = 0.5$ mm, respectively. The far field divergence θ_1 and θ_2 define the asymptote lines for the two beams, respectively. z_{R1} and z_{R2} of the two laser beams are marked in Fig. 2.5.

Figure 2.6 plots Eq. (2.2) for two laser beams same as the two beams in Fig. 2.5. It can be seen from Eq. (2.2) and Fig. 2.6 that at the beam waist, both beams have a

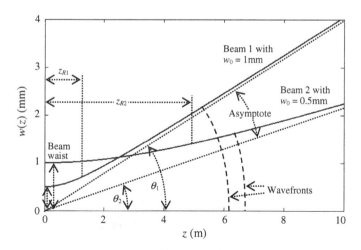

Fig. 2.5 The *solid curves* are $w(z)$ versus z for two laser beams with $w_0 = 1.0$ and 0.5 mm, respectively, both beams have $\lambda = 0.635\ \mu m$ and $M^2 = 1$

Fig. 2.6 $R(z)$ versus z for two laser beams with $w_0 = 1.0$ and 0.5 mm, respectively, $\lambda = 0.635$ μm and $M^2 = 1$

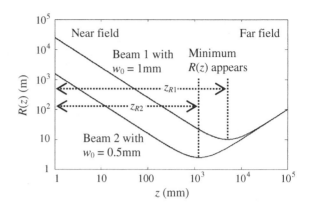

plane wavefront with radius $R(0)$ approaching infinity. As the beam propagates, $R(z)$ gradually decreases. The minimum $R(z)$ appears at $z = z_R$. As the beam continues propagating, the beam wavefront gradually becomes spherical, then $R(z)$ becomes proportional to z. z_R is often used as a criterion, $z \ll z_R$ is "near field", $z \gg z_R$ is "far field", and $z \sim z_R$ is the intermediate field.

Figure 2.7 plots Eq. (2.3) for a laser beam with Gaussian intensity distribution in an arbitrary cross section perpendicular to the propagation direction of the beam, where $I_0(z)$ is normalized to 1. Beam radius is usually defined at either $1/e^2$ intensity level or at FWHM level. We can find from Eq. (2.3) that the $1/e^2$ intensity radius equals $w(z)$, and the half magnitude radius equals $0.59w(z)$. The ratio between these two radii is about 1.7.

The percentage of laser energy encircled inside the $1/e^2$ intensity radius can be calculated by

$$\frac{\int_0^{w(z)} e^{-2r^2/w(z)^2} r dr}{\int_0^{\infty} e^{-2r^2/w(z)^2} r dr} = 86.4\,\% \qquad (2.6)$$

Fig. 2.7 Normalized Gaussian intensity distribution

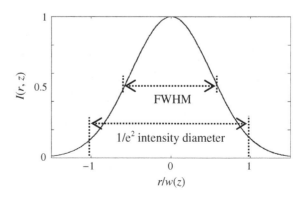

where r is the radial variable. Similarly, the percentage laser energy encircled inside the half magnitude radius can be calculated by

$$\frac{\int_0^{0.59w(z)} e^{-2r^2/w(z)^2} r \, dr}{\int_0^{\infty} e^{-2r^2/w(z)^2} r \, dr} = 69.2\,\%. \tag{2.7}$$

The characteristics of basic TE mode Gaussian beams have been studied extensively. Many works studying this subject have been published. The most cited one is probably the book *Lasers* written by Siegmann [5].

2.3.2 M^2 Factor Approximation

The beams of some solid state lasers and laser diodes are not exact basic mode Gaussian beams, they may contain higher order Gaussian modes. It is difficult to find the mode structure details in these beams, since the unavoidable measurement errors often lead to inconclusive results. A practical way of handling such laser beams is to neglect the mode structure details, assume the beams still have Gaussian intensity distributions, and introduce a M^2 factor to the beams [6, 7]. By definition, $M^2 = 1$ means the beam is a perfect basic mode Gaussian beam. $M^2 \geq 1$ means the beam deviates from a basic mode Gaussian beam.

Figures 2.8 and 2.9 plot Eqs. (2.1) and (2.2) for two beams with the same waist size and wavelength, but different $M^2 = 1$ and 1.2, respectively. We can see that the beam far field divergence is proportional to the value of the M^2 factor. Most collimated single TE mode laser diode beams have a M^2 of 1.1 and 1.2. The introduction of the M^2 factor enables the equation set for basic mode Gaussian beam to describe non-basic mode Gaussian with reasonable accuracy and thereby significantly simplify the mathematics involved. M^2 factor has been widely used now to describe various quasi-Gaussian laser beams. Some laser developers even use M^2 factor to

Fig. 2.8 *Solid curves* are w (z) versus z for two laser beams with $M^2 = 1$ and 1.2, respectively. Both beams have $w_0 = 1.0$ mm and $\lambda = 0.635$ μm

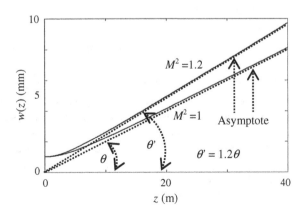

Fig. 2.9 *Solid curves* are R (z) versus z for two laser beams with $w_0 = 1.0$ mm, $\lambda = 0.635$ mm, and $M^2 = 1$ and 1.2, respectively

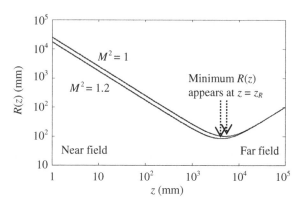

describe multi-TE mode laser beams. ISO has established a detail procedure for measuring M^2 factor [8].

2.3.3 Thin Lens Equation for a Real Laser Beam

Thin lens equation was originally derived as a simple analytical model to describe how a lens manipulates geometric rays. Thin lens equation is an approximated model, but accurate enough in most applications, and is therefore widely used. Thin lens equation has the form [9]

$$\frac{i}{f} = \frac{o}{o - f} \qquad (2.8)$$

where o is the object distance measured from the object point to the lens principal plane. The lens focuses the rays from the object point and produces an image of the object point, i is the image distance measured from the image point to the lens principal plane, and f is the focal length of the lens. The lens magnification ratio m is defined as

$$m = \frac{i}{o} \qquad (2.9)$$

Equation (2.8) shows that for $o = f_+$, $i \rightarrow \infty$ and $m \rightarrow \infty$, the rays are collimated, where f_+ means a value slightly larger than f. For $o = f_-$, $i \rightarrow -\infty$ and $m \rightarrow -\infty$, the rays are also collimated, where f_- means a value slightly smaller than f. For $o \rightarrow \infty$, $i \rightarrow f$ and $m \rightarrow 0$, the rays are focused. It is noted that the geometric optics is not accurate to calculate the size of a focused spot, the actual smallest possible focused spot radius is the diffraction limited radius $1.22\lambda f/d$, where d is the ray bundle diameter.

Equation (2.8) was first modified to be applicable to a basic mode Gaussian beam without considering the M^2 factor [10] and was later expanded to include the M^2 factor [4]. The latest form of thin lens equation looks like

$$\frac{i}{f} = \frac{\frac{o}{f}\left(\frac{o}{f} - 1\right) + \left(\frac{z_R}{f}\right)^2}{\left(\frac{o}{f} - 1\right)^2 + \left(\frac{z_R}{f}\right)^2} \qquad (2.10)$$

where o is the object distance measured from the waist of the laser beam incident on the lens to the principal plane of the lens, i is the image distance measured from the waist of the laser beam output from the lens to the principal plane of the lens, and z_R is the incoming Rayleigh range of the incident beam defined in Eq. (2.4). The M^2 factor is included in z_R. z_R/f is an important parameter in Eq. (2.10). For $z_R/f \to 0$, Eq. (2.10) reduces to Eq. (2.8), which means such a laser beam can be treated as geometric rays emitted by a point source. For $z_R/f \to \infty$, Eq. (2.10) leads to $i = f$, the laser beam is focused at the focal point of the lens.

Equation (2.10) has some interesting characteristics that are different from those of Eq. (2.8). One characteristics is the maximum and minimum focusing distance that can be found by differentiating Eq. (2.10) and assuming $\Delta i/\Delta o = 0$, we obtain

$$o = f \pm z_R \qquad (2.11)$$

Plugging $o = f + z_R$ into Eq. (2.10), we can find the maximum focusing distance to be

$$i_{max} = f \frac{\frac{2z_R}{f} + 1}{\frac{2z_R}{f}} \qquad (2.12)$$

Plugging $o = f - z_R$ into Eq. (2.10) we can find the minimum focusing distance to be

$$i_{min} = f \frac{\frac{2z_R}{f} - 1}{\frac{2z_R}{f}} \qquad (2.13)$$

z_R/f again plays an important role here. For $z_R/f \gg 1$, Eqs. (2.12) and (2.13) reduce to $i_{max} = i_{min} = f$, which is a focusing situation. For $z_R/f \ll 1$, Eqs. (2.12) and (2.13) reduce to $i_{max} \to \infty$ and $i_{min} \to -\infty$, the beam is collimated similar to collimated geometric rays emitted by a point source.

For a typical laser diode beam, z_R is several microns; assuming this laser diode beam is collimated by a lens with a focal length of several millimeters, we have $z_R/f \sim 0.001$, Eq. (2.12) reduces to $i_{max} \approx f^2/2z_R \approx 500f \sim 1$ m, and Eq. (2.13) reduces to $i_{min} \approx -f^2/2z_R \approx -500f \sim -1$ m. The negative value of i_{min} indicates that the laser beam outgoing from the lens has an imaginary waist on the left-hand side of the collimation lens.

The waist of a collimated laser diode beam is a few millimeters, the z_R of such a collimated beam is several meters. When this collimated laser diode beam is focused by a lens with a focal length of several millimeters, we have $z_R/f \sim 1000$, Eqs. (2.12) and (2.13) give $i_{max} \approx 1.001f$ and $i_{min} \approx 0.999f$, respectively. This means the position of the focused spot of the beam can shift around the lens focal point in the range of ~ 1 µm.

Equations (2.10) and (2.8) are plotted in Fig. 2.10 by solid and dashed curves, respectively, with $z_R/f = 0.1, 0.2, 0.4$ and 1, respectively. The maximum and minimum focusing distances i_{max} and i_{min} are marked by the open circles and open squares on each curve. Figure 2.10 shows that for $o/f = 1$, $i/f = 1$ for any z_R/f values. For smaller z_R/f, i changes faster as o changes and the values of i_{max} and $|i_{min}|$ are larger. When $z_R/f \rightarrow 0$, the Gaussian beam reduces to a point source, the solid curve approaches the dashed curve.

The magnification of a lens on a laser beam propagating through the lens is defined by the ratio of w_0'/w_0, where w_0' is the waist radius of the beam output from the lens. w_0'/w_0 can be found by modifying Eq. (2.9) as [4].

$$m = \frac{w_0'}{w_0}$$

$$= \frac{1}{\left[\left(\frac{o}{f} - 1 \right)^2 + \left(\frac{z_R}{f} \right)^2 \right]^{0.5}} \tag{2.14}$$

The M^2 factor is included in z_R as defined in Eq. (2.4). $w_0'/w_0 \gg 1$ indicates the beam is collimated. $w_0'/w_0 \ll 1$ indicates the beam is focused. $w_0'/w_0 \sim 1$ means

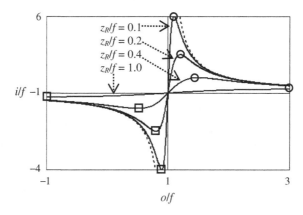

Fig. 2.10 The *solid curves* are i/f versus o/f curves with $z_R/f = 0.1, 0.2, 0.4$ and 1, respectively. The *dashed curves* plotted here for comparison are for geometric rays emitted by a point source. i_{max} and i_{min} are marked by the *open circles* and *open squares* on each curve

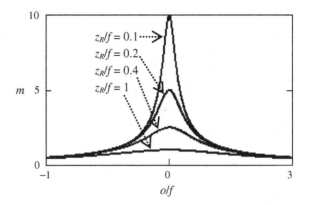

Fig. 2.11 Lens magnification m versus o/f curve with z_R/f being a parameter

the beam is relayed. From Eq. (2.14) we can see that $w_0'/w_0 = 1$ can appear for various combinations of o/f and z_R/f. It can be seen that for $z_R/f \rightarrow 0$, the Gaussian beam reduces to a point source and Eq. (2.14) reduces to Eq. (2.9).

Equation (2.14) is plotted in Fig. 2.11 with z_R/f being a parameter. We can see from Fig. 2.11 that for any z_R/f values, m peaks at $o/f = 1$. For $o/f = 1$, m is larger for smaller z_R/f, since this is a collimation situation, a smaller waist size incident beam means larger divergence and larger waist size of collimated beam. For $z_R/f > 1$, the value of m does not change much as the value of o/f changes, since this is a focusing situation; the waist size of the focused spot does not change much when the incident beam waist location changes.

2.3.4 Non-paraxial Gaussian Beams

Laser beams with larger divergent angles can be non-paraxial Gaussian beams and cannot be treated accurately by the basic mode paraxial Gaussian model.

Nemoto [11] shows that when $ks_0 < 4$ the paraxial Gaussian model deviates appreciably from the exact solution and that when $ks_0 < 2$ the paraxial Gaussian model differs considerably from the exact solution, where s_0 is the $1/e$ intensity radius of the beam waist and $k = 2\pi/\lambda$ is the wave vector. s_0 can be converted to the more commonly used $1/e^2$ intensity radius w_0 of the beam waist by $s_0 = 0.368w_0$, then Nemoto's two conditions can be written, respectively, as

$$w_0 < 1.73\lambda \tag{2.15}$$

$$w_0 < 0.87\lambda \tag{2.16}$$

It would be difficult to directly measure the waist radius w_0 of an un-manipulated beam of laser diode, since the waist is at the emission facet and is likely only about 1 μm, but it would be much easier to measure the far field divergence of the un-manipulated beam. Also, every datasheet of laser diode provides the far field FWHM divergence $2\theta_{FWHM}$, not w_0. It will be more convenient to replace w_0 in Eqs. (2.15) and (2.16) by the far field divergence. The paraxial Gaussian model relates the $1/e^2$ intensity far field half divergence θ to w_0 by Eq. (2.5). We know that Eq. (2.5) itself is a paraxial approximation and we are now talking about the inaccuracy of paraxial Gaussian model. But it is still adequate to use Eq. (2.5) to provide a criterion for assessment. Combining Eqs. (2.5), (2.15), and (2.16) to eliminate w_0 and converting θ from radian to degree, we obtain two conditions in terms of degree

$$2\theta > 21° \quad \text{or} \quad 2\theta_{FWHM} > 12.4° \qquad (2.17)$$

for paraxial Gaussian model deviates appreciably from the exact solution and

$$2\theta > 42° \quad \text{or} \quad 2\theta_{FWHM} > 24.7° \qquad (2.18)$$

for paraxial Gaussian model considerably differs from the exact solution. Checking the datasheets of various laser diodes, we can find that the slow axis divergence of almost all laser diodes does not meet these two conditions, paraxial Gaussian model is accurate enough to treat laser diode beams in the slow axis direction, and that the fast axis divergence of many laser diodes meets Eq. (2.17) or even Eq. (2.18), which means many laser diode beam are non-paraxial in the fast axis direction. As we will show later in Sect. 3.2, if we can accept an error of 10 % or so, then the paraxial Gaussian model discussed in this chapter can still be used to treat laser diode beams in the fast axis direction. Otherwise, we have to use Kirchhoff diffraction integral to perform accurate numerical analysis.

Figure 2.12 shows the far field angular intensity distribution of a non-paraxial laser diode beam at 5 mm. The solid curve is accurate obtained using Kirchhoff

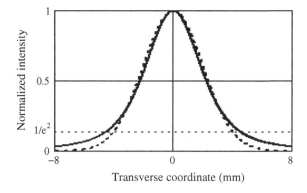

Fig. 2.12 Normalized intensity profiles at 5 mm for a non-paraxial Gaussian beam with a $1/e^2$ intensity radius of 0.25 μm and a wavelength of 0.635 μm. *Solid curve* is obtained using Kirchhoff diffraction integration. *Dashed curve* is obtained using paraxial Gaussian model

diffraction integration. The dashed curve is approximation obtained using paraxial Gaussian model. We will discuss the characteristics of non-paraxial Gaussian beam in detail in Sect. 3.2.

2.3.5 Raytracing Technique

2.3.5.1 ABCD Matrix Method

A geometric ray propagating through optical elements can be conveniently analyzed by ABCD matrix method [12]. Below we consider a simple example. As shown in Fig. 2.13a, a ray propagates from a medium with index n_1 to another medium with index n_2. The interface of these two media is planar. The input ray can be described by its height x_1 when it hits the optics surface and its angle θ_1 to the optical axis. Similarly, the output ray can be described by its height x_2 when it leaves the optics surface and its angle θ_2 to the optical axis. We have

$$x_2 = x_1 \tag{2.19}$$

$$\theta_2 = \frac{n_1}{n_2}\theta_1 \tag{2.20}$$

Equation (2.20) is the paraxial form of Snell's law [13]. We can write Eqs. (2.19) and (2.20) in the matrix form

$$\begin{bmatrix} x_2 \\ \theta_2 \end{bmatrix} = \begin{bmatrix} A & B \\ C & D \end{bmatrix} \begin{bmatrix} x_1 \\ \theta \end{bmatrix}$$
$$= \begin{bmatrix} 1 & 0 \\ 0 & \frac{n_1}{n_2} \end{bmatrix} \begin{bmatrix} x_1 \\ \theta_1 \end{bmatrix} \tag{2.21}$$

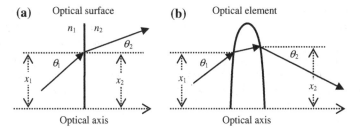

Fig. 2.13 **a** A geometric ray propagates through an optical surface. **b** A geometric ray propagates through a lens

Table 2.1 Commonly used ABCD matrix

Optical element	Matrix
Propagation in a uniform medium	$\begin{bmatrix} 1 & d \\ 0 & 1 \end{bmatrix}$
Refraction at a planar surface	$\begin{bmatrix} 1 & 0 \\ 0 & \dfrac{n_1}{n_2} \end{bmatrix}$
Refraction at a curved optical surface	$\begin{bmatrix} 1 & 0 \\ \dfrac{n_1 - n_2}{R \cdot n_2} & \dfrac{n_1}{n_2} \end{bmatrix}$
Reflection from a planar mirror	$\begin{bmatrix} 1 & 0 \\ 0 & 1 \end{bmatrix}$
Reflection from a curved mirror	$\begin{bmatrix} 1 & 0 \\ \dfrac{2}{R} & 1 \end{bmatrix}$
Thin lens	$\begin{bmatrix} 1 & 0 \\ -\dfrac{1}{f} & 1 \end{bmatrix}$
Thick lens	$\begin{bmatrix} 1 & 0 \\ \dfrac{n_2 - n_1}{R_2 \cdot n_1} & \dfrac{n_1}{n_2} \end{bmatrix} \begin{bmatrix} 1 & t \\ 0 & 1 \end{bmatrix} \begin{bmatrix} 1 & 0 \\ \dfrac{n_1 - n_2}{R_1 \cdot n_2} & \dfrac{n_1}{n_2} \end{bmatrix}$

Equation (2.21) shows that an optical surface can be described by a 2 × 2 matrix. Figure 2.13b shows a more general case, a ray propagates through a lens, then x_2 is not necessary equal to x_1. If a geometric ray propagates through n optical elements, the height and angle of the output ray can be calculated by

$$\begin{bmatrix} x_n \\ \theta_n \end{bmatrix} = \begin{bmatrix} A_1 & B_1 \\ C_1 & D_1 \end{bmatrix} \cdots \begin{bmatrix} A_n & B_n \\ C_n & D_n \end{bmatrix} \begin{bmatrix} x_1 \\ \theta_1 \end{bmatrix} \qquad (2.22)$$

Each matrix in Eq. (2.22) describes one optical surface. The process of solving Eq. (2.22) is much simpler than the process of exhaustively tracing the ray through every optical surface.

Reference [12] and many other optics text books provide a list of matrices for various commonly used optical elements. For readers' convenience, we re-produce with minor modifications a list here in Table 2.1. It is not difficult to prove these matrices. In the table, d is the axial distance, R is the radius of curvature, $R > 0$ for convex surface and $R < 0$ for concave surface, f is the focal length, $f > 0$ for positive lens and $f < 0$ for negative lens, and n_1 and n_2 are the initial and final refractive indexes, respectively. For the thick lens, t is the lens center thickness, n_1 and n_2 are the refractive indexes outside and inside the lens, respectively, and R_1 and R_2 are the radii of curvature of the first and second surfaces, respectively.

2.3.5.2 Apply ABCD Matrix to a Gaussian Beam

The ABCD matrix method was originally developed to analyze geometric rays propagating through optical elements. These rays in a uniform medium are straight lines. By definition, a ray propagates in the direction of wavefront normal. For a Gaussian beam, the wavefront radius and wavefront center positions change as the beam propagates, therefore the propagation direction of a "ray" in a Gaussian beam also changes.

To apply ABCD matrix method to analyze the propagation of a Gaussian beam, we need to conceive a ray in the beam and follow this ray through the optical element. Considering an example of a thin lens shown in Fig. 2.14, we can conceive an input ray and an output ray for the beam, the rays are the tangents of any intensity contours of the input and output beams at the lens, respectively. It is more convenient to conceive the rays at the $1/e^2$ intensity level. Because we already know the input beam data, we can find the $1/e^2$ intensity height $w(z)$ and the $1/e^2$ intensity divergent angle θ for the input ray, as shown in Fig. 2.14. Note that the lens is not necessary at the far field of the beam, θ here is not necessary for the far field divergent angle given by Eq. (2.5). Applying the ABCD matrix to the input ray, we have

$$\begin{bmatrix} w'(z) \\ \theta' \end{bmatrix} = \begin{bmatrix} 1 & 0 \\ -\frac{1}{f} & 1 \end{bmatrix} \begin{bmatrix} w(z) \\ \theta \end{bmatrix} \tag{2.23}$$

where $w'(z)$ and θ' are the $1/e^2$ intensity height and divergent angle for the output ray.

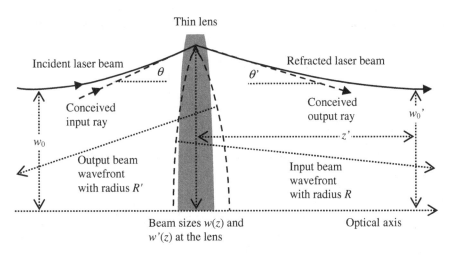

Fig. 2.14 A laser beam propagates through a thin lens

Solving Eq. (2.23), we obtain

$$w'(z) = w(z) \tag{2.24}$$

$$\theta' = -\frac{w(z)}{f} + \theta \tag{2.25}$$

Equation (2.24) is obvious, as can be seen in Fig. 2.14. We also have paraxial relations $\theta = w(z)/R$ and $\theta' = w'(z)/R'$, where R and R' are the wavefront radii of the input and output rays at the lens, respectively, as shown in Fig. 2.14, and R is known. Inserting the two relations into Eq. (2.25), we obtain R'

$$\frac{1}{R'} = \frac{1}{R} - \frac{1}{f} \tag{2.26}$$

Equation (2.26) is the same as the geometric thin lens equation.

Having obtained $w'(z)$ and R', we can back calculate the $1/e^2$ intensity waist radius w_0' and waist location z' of the output beam by modifying Eqs. (2.1) and (2.2) to

$$w_0' = \frac{w'(x)}{\left[1 + \frac{w'(z)^4 \pi^2}{R'(z)^2 (M^2 \lambda)^2}\right]^{0.5}} \tag{2.27}$$

$$z' = \frac{R'(x)}{1 + \frac{R'(z)^2 (M^2 \lambda)^2}{w'(z)^4 \pi^2}} \tag{2.28}$$

The derivation of Eqs. (2.27) and (2.28) is a little complex, we write the main steps here. Equations (2.1) and (2.2) are rewritten here for readers' convenience.

$$w'(z) = w_0' \left[1 + \left(\frac{M^2 \lambda z'}{\pi w_0'^2}\right)\right]^{1/2} \tag{2.1}$$

$$R'(z) = z' \left[1 + \left(\frac{\pi w_0'^2}{M^2 \lambda z'}\right)\right] \tag{2.2}$$

Taking the square of both sides of Eq. (2.1) and dividing the result by Eq. (2.2), we obtain

$$\frac{w'(z)^2}{R'(z)} = \frac{w_0'^2 \left[1 + \left(\frac{M^2 \lambda z'}{\pi w_0'^2}\right)^2\right]}{z' \left[1 + \left(\frac{\pi w_0'^2}{M^2 \lambda z'}\right)^2\right]} \tag{2.29}$$

Taking the $(M^2\lambda z')^2/(\pi w_0'^2)^2$ term out of the parenthesis in the numerator of Eq. (2.29) and canceling the same terms in the numerator and denominator, Eq. (2.29) becomes

$$\frac{w'(z)^2}{R'(z)} = \frac{w_0'^2 \left(\frac{M^2\lambda z'}{\pi w_0'^2}\right)^2 \left[1 + \left(\frac{\pi w_0'^2}{M^2\lambda z'}\right)^2\right]}{z'\left[1 + \left(\frac{\pi w_0'^2}{M^2\lambda z'}\right)^2\right]}$$

$$= \frac{z'}{w_0'^2}\left(\frac{M^2\lambda}{\pi}\right)^2 \quad \text{or}$$

$$\frac{z'}{w_0'^2} = \frac{w'(z)^2}{R'(z)}\left(\frac{\pi}{M^2\lambda}\right)^2$$

(2.30)

Inserting Eq. (2.30) into the parentheses of Eqs. (2.1) and (2.2), and solving for w_0' and z', respectively, we obtain Eqs. (2.27) and (2.28).

Reference [14] provides a detailed study on ray equivalent modeling of Gaussian beams.

2.4 Zemax Modeling of a Gaussian Beam Propagating Through a Lens

Zemax is probably the most widely used optical design software. Zemax can perform sequential raytracing for designing imaging optics and non-sequential raytracing for designing illumination optics. Zemax offers three editions with different capabilities and prices. The two higher editions, Professional and Premium editions, have the feature of modeling Gaussian beams propagating through optics, a useful tool that can save a lot time and effort when designing laser optics. Although in its 2014 manual, Zemax describes this feature in the *Physical Optics* section in Chapter 7, Analysis Menu and in Chapter 26, Physical Optics Propagation, this feature is still not well known to many users. In this section, we use two examples to demonstrate, step-by-step, how to use Zemax to model a Gaussian beam propagating through a lens. We assume the readers are already familiar with the geometric raytracing feature of Zemax and will emphasize the procedure difference between Zemax sequential raytracing and Gaussian beam modeling.

We use mm as the length unit throughout Sect. 2.4.

2.4.1 Collimating a Gaussian Beam

To start designing optics using Zemax sequential raytracing, the first three parameters to be selected are the *Field, General/Aperture,* and *Wavelength* in the

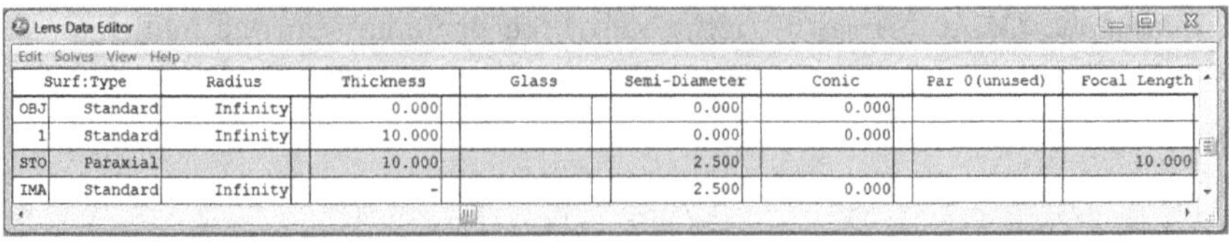

Surf:Type		Radius	Thickness	Glass	Semi-Diameter	Conic	Par 0(unused)	Focal Length
OBJ	Standard	Infinity	0.000		0.000	0.000		
1	Standard	Infinity	10.000		0.000	0.000		
STO	Paraxial		10.000		2.500			10.000
IMA	Standard	Infinity	-		2.500	0.000		

Fig. 2.15 Zemax *Lens Data Editor* for modeling a Gaussian beam propagating through a paraxial lens

System drop-down list. The first two parameters are irrelevant for modeling Gaussian beam, we randomly pick up 0° field and 5 mm aperture. We still need select wavelength since it is a parameter of Gaussian beam. Here we pick up 0.65 μm for the wavelength, just for demonstration purpose. Then we open the *Lens Data Editor* box. The thickness of the *OBJ* surface is important for sequential raytracing, but irrelevant again for modeling Gaussian beam, we type in 0 for simplicity. We select *Surface 2* as the *STP* surface and place an ideal *Paraxial* lens with 10 mm focal length at the *STO* surface. We type in 10 in the *Surface 1 Thickness* box, *Surface 1* is then 10 mm away from the *STO* surface and is at the focal plane of the lens. Type in 10 in the surface *STO Thickness* box, which means surface *IMA* is 10 mm away from surface *STO* and is at another focal plane of the lens. After typing in these data, the *Lens Data Editor* box will look like as shown in Fig. 2.15.

Then we need to set the parameters for the Gaussian beam to be modeled. Click *Analysis* button, in the drop-down list, click *Physical optics* button, there are four choices in the drop-down list: *Paraxial Gaussian beam*, *Skew Gaussian beam*, *Physical Optics Propagation*, and *Beam File Viewer*. Laser diode beams are skew (elliptical) Gaussian beams. But here we only model a paraxial (circular) Gaussian beam for simplicity. After we have mastered the modeling process, we can model the skew Gaussian beam feature without difficulties.

Click the *Paraxial Gaussian beam* button, a text box *Paraxial Gaussian Beam Data* appears, we will look at it later. Click the *Settings* button at the top of the text box, a table box *Paraxial Gaussian Beam Settings* appears. Then we can type in the parameters of the laser diode beam. In the table box, we have only one choice for the wavelength, that is, 0.65 μm selected by us earlier. Type 0.002 in *Waist size* box, which means the $1/e^2$ intensity waist (radius) of the embedded fundamental mode beam of the input beam is 0.002 mm. This small waist size is common for laser diodes and such a small beam has large divergence. An ideal Gaussian beam has $M^2 = 1$, here we type 1.2 in the M^2 *factor* box for a mixed mode beam. When modeling Gaussian beam, the object is the beam waist, the distance between the beam waist and surface 1 is defined by the value in the *Surf 1 to waist* box. When we type in 0, the beam waist is set at surface 1 and is at the focal plane of the paraxial lens, a negative value here means the beam waist is at the left side of the surface. Since the Rayleigh range of this beam is much smaller than the focal length of the lens, this is a collimating situation, as explained in Sect. 3.1.4. For a skew Gaussian beam, the beam behaves differently in the *x-z* and *y-z* planes. For a

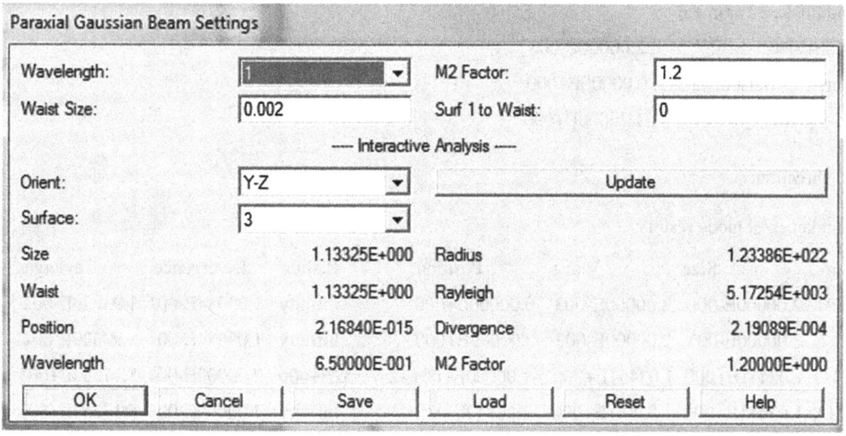

Fig. 2.16 Content of table box *Paraxial Beam Settings* after typing in all the data and clicking *Update*

circular Gaussian beam modeled here, the orientation does not matter, we randomly choose *Y-Z* in the *Orient* box. The number filled in the *Surface* box selects the surface at which the beam parameters will be shown in the table box; we select number 3 which is the last surface. We will see soon that the beam parameters at every surface will be shown later in the text box *Paraxial Gaussian Beam Data*. After typing in all these numbers, click the *Update* button in the table box, the table box will look like as shown in Fig. 2.16. The beam data shown in the lower half of the box is for surface 3. We do not explain these data now, since they will be shown in the text box again.

Click the *Update* button in the text box *Paraxial Gaussian Beam Data*, the relevant part of the text box will look like as shown in Fig. 2.17. In Zemax sequential raytracing, any one surface can be selected as *Global coordinate reference*, the positions of all other surfaces are relative to this surface. In Zemax Gaussian beam modeling, the beam waist position is relative to any surface that is under consideration. We need to keep this difference in mind when interpreting the modeling results shown in Fig. 2.17. We also note that all the beam data shown for a surface is AFTER the beam propagating through the surface.

Let us first check the beam data at every surface for the *Fundamental mode results* shown in Fig. 2.17. This results are for an ideal Gaussian beam with $M^2 = 1$ embedded in the beam we set with $M^2 = 1.2$.

OBJ surface. Since we let the distance between *OBJ* surface and surface *1* be 0, all the beam data in these two surfaces are the same.
Surface *1*:

(Beam) *Size* at this surface is 2.00000E−3. Because the beam waist we typed in is 0.002 mm and the waist is at this surface.
Waist (size) is 2.00000E−3, as we typed in earlier.

Input Beam Parameters:

Waist size   : 2.00000E-003

Surf 1 to waist distance : 0.00000E+000

M Squared   : 1.20000E+000

Y-Direction:

Fundamental mode results:

Sur	Size	Waist	Position	Radius	Divergence	Rayleigh
OBJ	2.00000E-000	2.00000E-003	0.00000E+000	Infinity	1.03084E-001	1.93329E-002
1	2.00000E-003	2.00000E-003	0.00000E+000	Infinity	1.03084E-001	1.93329E-002
STO	1.03451E+000	1.03451E+000	-1.00000E+001	-2.67552E+006	2.00000E-004	5.17254E+003
IMA	1.03451E+000	1.03451E+000	2.16840E-015	Infinity	2.00000E-004	5.17254E+003

Mixed Mode results for M2 = 1.2000:

Sur	Size	Waist	Position	Radius	Divergence	Rayleigh
OBJ	2.19089E-003	2.19089E-003	0.00000E+000	1.00000E+010	1.12923E-001	1.93329E-002
1	2.19089E-003	2.19089E-003	0.00000E+000	1.00000E+010	1.12923E-001	1.93329E-002
STO	1.13325E+000	1.13325E+000	-1.00000E+001	-2.67552E+006	2.19089E-004	5.17254E+003
IMA	1.13325E+000	1.13325E+000	2.16840E-015	1.23386E+022	2.19089E-004	5.17254E+003

Fig. 2.17 Zemax modeling of a paraxial lens collimating a Gaussian beam. Shown here is the content of the text box *Paraxial Gaussian beam Parameters*. Only the *Y-direction* content is pasted here, since the beam is circular; the *X-direction* content is the same and is neglected

(Beam waist) *Position* is 0.00000E+000, which means the distance between the beam waist and surface 1 is 0. This is what we typed in earlier.

Radius is infinity. Since the beam waist is at surface 1, the wavefront at surface 1 must be flat.

Divergence is 1.03084E−001. Note that the divergence is the far field divergence, not the divergence at this specific surface. For a given waist size, wavelength, and M^2 factor, there is only one far field divergence as given by Eq. (2.5), no matter what surface we are considering. Zemax calculated the divergence based on the beam parameters we type in.

Rayleigh (range) is 1.93329E–002. Again, for a given waist size, wavelength, and M^2 factor, there is only one Rayleigh range given by Eq. (2.4), no matter what surface we are considering. Zemax calculated the Rayleigh range based on the beam parameters we type in.

Surface *STO*

Since we put a paraxial lens at surface *STO* and the waist of the input beam is at the focal plane of the lens, the beam is collimated after it passes through surface *STO*.

(Beam) *Size* is 1.03451E+000 calculated by Zemax.

Waist (size) is 1.03451E+000 calculated by Zemax. Note that the waist of the collimated beam is at surface *IMA* as will be shown in the *Position* below, but the waist size and beam size at surface *STO* are virtually the same, because the beam is collimated and the two surfaces are only 10 mm apart.

(Beam waist) *Position* is −1.00000E+001, which means the waist of the collimated beam is 10 mm away from surface *STO*. The negative sign here indicates that the waist of the collimated beam is on the right side of surface *STO*.

Radius is −2.67552E+006, surface *STO* is only 10 mm away from surface *IMA* where the waist of the collimated beam is located. The wavefront radius at surface *STO* must be large.

Divergence is 2.00000E−4, very small, since the beam is collimated with a waist size of about 1 mm.

Rayleigh is 5.17254E+003 and should match the waist size of the collimated beam.

Surface *IMA*

The *Size* and *Waist* are the same as those at surface *STO*, since the beam is collimated and these two surfaces are only 10 mm apart. The beam waist *Position* is relative to surface *IMA* and is virtually 0. The positive sign means the beam waist position is slightly on the left-hand side of surface *IMA*. The *Radius* is infinity because the beam waist is at this surface.

Now, let us check the *Mixed Mode results for* M^2 = 1.2000 in Fig. 2.17. The *Position*, *Radius*, and *Rayleigh* are virtually the same as those for the *Fundamental mode*. The *Waist* of the input beam is about 1.2 times larger than the *Waist* of the embedded beam because we select M^2 = 1.2. The *Waist* and *Size* of the collimated beam are about 1.1 times larger than those of the embedded beam. The *Divergence* is about 1.1 times larger for every surface. The situation is illustrated in Fig. 2.18.

We note here that Eq. (2.5) shows that the far field divergence is proportional to the value of the M^2 factor and inversely proportional to the beam waist size. In Fig. 2.17, the mixed mode has a waist radius 10 % larger and an M^2 factor value

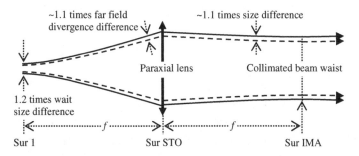

Fig. 2.18 The *solid curves* are for a Gaussian beam with M^2 = 1.2, the *dashed curves* are for the embedded Gaussian beam with M^2 = 1. The drawing is not to exact proportion for clarity

20 % larger than those of the fundamental mode. Therefore, the mixed mode has 10 % larger far field divergence than that of the fundamental mode.

2.4.2 Focusing a Gaussian Beam

Now we consider using the same *Paraxial* lens to focus a Gaussian beam.

In the table box *Paraxial Gaussian Beam Settings*, keep everything the same, only change the *Waist* value from 0.002 to 1. The Rayleigh range of a 1 mm waist size beam is much larger than the 10 mm focal length. The waist of such a beam at the focal plane of the lens means focusing. Then click the *Update* button at the text box *Paraxial Gaussian Beam Data*, the text box will look like as shown in Fig. 2.19. These numbers can be explained in the same way as in Sect. 2.4.1, we do not repeat it here. We can see that the *Waist* at surface *IMA* is only about 2 μm because it is focusing.

If we want to see the beam data at any other location, we can simply insert new surfaces into the *Lens Data Editor* at these locations. We can also type in other lens

Input Beam Parameters:

Waist size   : 1.00000E+000

Surf 1 to waist distance: 0.00000E+000

M Squared   : 1.20000E+000

Y-Direction:

Fundamental mode results:

Sur	Size	Waist	Position	Radius	Divergence	Rayleigh
OBJ	1.00000E+000	1.00000E+000	0.00000E+000	Infinity	2.06901E-004	4.83322E+003
1	1.00000E+000	1.00000E+000	0.00000E+000	Infinity	2.06901E-004	4.83322E+003
STO	1.00000E+000	2.06901E-003	-1.00000E+001	-1.00000E+001	9.96687E-002	2.06901E-002
IMA	2.06901E-003	2.06901E-003	0.00000E+000	Infinity	9.96687E-002	2.06901E-002

Mixed Mode results for M2 = 1.2000:

Sur	Size	Waist	Position	Radius	Divergence	Rayleigh
OBJ	1.09545E+000	1.09545E+000	0.00000E+000	1.00000E+010	2.26649E-004	4.83322E+003
1	1.09545E+000	1.09545E+000	0.00000E+000	1.00000E+010	2.26649E-004	4.83322E+003
STO	1.09545E+000	2.26649E-003	-1.00000E+001	-1.00000E+001	1.09182E-001	2.06901E-002
IMA	2.26649E-003	2.26649E-003	0.00000E+000	1.00000E+010	1.09182E-001	2.06901E-002

Fig. 2.19 Zemax modeling of a Paraxial lens focusing a Gaussian beam. Content of the text box *Paraxial Gaussian beam data*

data and beam data to model other propagations. If a real lens is used, the lens aberration must be well corrected. Strong aberrations will deviate a Gaussian beam from basic mode and the result of modeling such a beam is not accurate. We also note that the meaning of signs "+" and "−" can be confusing and require full attention.

References

1. Naqwi, A., et al.: Focusing of diode laser beams: a simple mathematical model. Appl. Opt. **29**, 1780–1785 (1990)
2. Li, Y.: Focusing of diode laser beams: a simple mathematical model: comment. Appl. Opt. **31**, 3392–3393 (1992)
3. Sun, H.: Modeling the near field and far field modes of single spatial mode laser diodes. Opt. Eng. **51**, 044202 (2012)
4. Sun, H.: Thin lens equation for a real laser beam with weak lens aperture truncation. Opt. Eng. **37**, 2906–2913 (1998)
5. Siegmann, A.E.: Lasers, Chapter 16 Wave Optics and Gaussian Beams and Chapter 17 Physical Properties of Gaussian Beams. University Science Books, Mill Valley (1986)
6. Siegman, A.E.: New developments in laser resonators. Proc. SPIE **1224**, 2–14 (1990)
7. Siegman, A.E.: Defining, measuring, and optimizing laser beam quality. Proc. SPIE **1868**, 2 (1993)
8. ISO Standard 11146: Lasers and laser-related equipment—test methods for laser beam widths, divergence angles and beam propagation ratios (2005)
9. Almost any optics text book discusses the thin lens equation
10. Self, S.A.: Focusing of spherical Gaussian beams. Appl. Opt. **22**, 658–661 (1983)
11. Nemoto, S.: Nonparaxial Gaussian beams. Appl. Opt. **29**, 1940–1946 (1990)
12. http://en.wikipedia.org/wiki/Ray_transfer_matrix_analysis
13. Almost any optics text book discusses Snell's law
14. Herloski, R., Marshall, S., Antos, R.: Gaussian beam ray—equivalent modeling and optical design. Appl. Opt. **22**, 1168–1174 (1983)

Chapter 3
Laser Diode Beam Manipulations

Abstract Various techniques of manipulating laser diode beams are discussed. The emphases are on focusing, collimating, delivery, circularizing, astigmatism correction, and single mode fiber coupling of single TE mode laser diode beams. The manipulation of multi-TE mode laser diode beams is briefly discussed.

Keywords Aperture truncation · Astigmatism correction · Beam shape · Circularizing · Collimating · Fast axis · Fiber coupling · Focus · Gaussian · Lens · Multi-TE mode · Single TE mode · Slow axis · Spot size

We have discussed the basic characteristics of laser diode beams in the previous chapters. In this chapter we will discuss the optics for manipulating laser diode beams.

Some equations discussed in Chap. 2 will be used in this chapter. For the convenience of the readers, we rewrite these equations here:

$$w(z) = w_0 \left[1 + \left(\frac{M^2 \lambda z}{\pi w_0{}^2} \right)^2 \right]^{1/2} \tag{3.1}$$

$$R(z) = z \left[1 + \left(\frac{\pi w_0{}^2}{M^2 \lambda z} \right)^2 \right] \tag{3.2}$$

$$z_R = \frac{\pi w_0{}^2}{M^2 \lambda} \tag{3.3}$$

$$\theta = \frac{w(z)}{z}$$
$$= \frac{M^2 \lambda}{\pi w_0}$$
$$= \frac{w_0}{z_R} \tag{3.4}$$

© The Author(s) 2015
H. Sun, *A Practical Guide to Handling Laser Diode Beams*,
SpringerBriefs in Physics, DOI 10.1007/978-94-017-9783-2_3

$$\frac{i}{f} = \frac{\frac{o}{f}\left(\frac{o}{f} - 1\right) + \left(\frac{z_R}{f}\right)^2}{\left(\frac{o}{f} - 1\right)^2 + \left(\frac{z_R}{f}\right)^2} \tag{3.5}$$

$$m = \frac{w_0'}{w_0}$$

$$= \frac{1}{\left[\left(\frac{o}{f} - 1\right)^2 + \left(\frac{z_R}{f}\right)^2\right]^{0.5}} \tag{3.6}$$

$$i_{max} = f\frac{\frac{2z_R}{f} + 1}{\frac{2z_R}{f}} \tag{3.7}$$

where $w(z)$ is the $1/e^2$ intensity radius of a Gaussian beam at distance z measured from the beam waist, w_0 is the $1/e^2$ intensity radius of the beam waist, M^2 is the M square factor describing how much the beam deviates from basic mode Gaussian beam, λ is the wavelength, z_R is the Rayleigh range, f is the focal length of the collimating or focusing lens, i is the image distance defined as the distance between the lens principle plane and the waist position of the beam collimated or focused by the lens, o is the object distance defined as the distance between the lens principle plane and the waist position of the beam incident on the lens, m is the lens magnification, w_0' is the $1/e^2$ intensity radius of the waist of the collimated or focused beam, and i_{max} is the maximum image distance.

3.1 Collimating and Focusing

3.1.1 Lenses

3.1.1.1 Aspheric Lenses

Single element aspheric lenses are often used to collimate or focus laser diode beams, because they are smaller and lighter in weight than spherical lens groups. Tens of different types of aspheric lenses specially designed and fabricated for laser diode beam collimation are available in the market. These special aspheric lenses are of high quality and expensive and usually cost over 50 dollars per lens. Once a laser diode beam is collimated, it is no longer highly divergent and is much easier to handle; conventional spherical lenses can be used to further manipulate the collimated beams.

There are fewer aspheric lenses specially designed and fabricated for directly focusing a laser diode beam. Using an aspheric lens designed for collimating to focus a laser diode beam will cause severe aberration; the focused spot will be much

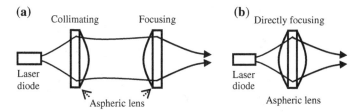

Fig. 3.1 **a** A commonly seen situation of using the first aspheric lens to collimate and second aspheric lenses to focus a laser diode beam. **b** A rarely seen situation of directly focusing a laser diode beam using one aspheric lens

larger than it should be. The most common way of focusing a laser diode beam is to combine two collimating lenses; the first lens is to collimate the beam and the second lens is to focus the collimated beam, as shown in Fig. 3.1a. Thereby, both lenses are used in their best conditions and the focused beam has the least aberrations. By properly choosing the focal length ratio of the two lenses used, we can obtain the desired magnification ratio or focused spot size. In the situation shown in Fig. 3.1a, the lens orientation matters. For most lenses, the flatter surface should face the laser diode or the focused spot.

Most aspheric lenses available in the market for collimating laser diode beams have a numerical aperture (NA) below 0.6. These lenses will more or less truncate the beam in the fast axis direction and cause side lobes, intensity ripples and focal shift, etc. In most applications, the beam truncation effects are acceptable or negligible. Further increasing the NA of a lens requires the use of a multielement lens group, such lens groups are rarely seen. The effects of aperture truncation on a beam will be discussed in Sects. 3.2.3 "Tightly focusing an astigmatic laser diode beam to a small spot" and 3.5 "Aperture beam truncation effects."

3.1.1.2 Lens Groups

Most aspheric lenses available in the market have an NA ≤ 0.6 and they do not have any color correction because only one type of glass is used. As we know, laser diodes of the same type can have a few nm wavelength tolerance and the same laser diode can have a few nm wavelength shift caused by temperature change. It can be shown that the single element aspheric lenses available in the market cannot maintain diffraction-limited performance over a wavelength range of ≥ 10 nm. In some high-demanding applications, we may need to use a lens group for larger NA and larger wavelength range. Figure 3.2 shows a high-performance lens group that contains six spherical lenses. The fabrication cost of these small lenses is about $100 per lens for a few pieces and about $20 per lens for large quantity, such as 100 pieces.

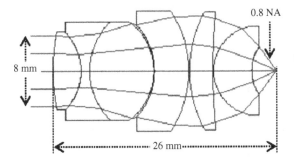

Fig. 3.2 A high-performance spherical lens groups for collimating or focusing laser diode beams. This lens group has a focal length of 5 mm, a NA of 0.8 and a diffraction-limited performance with Strehl ratio of 0.998 over a 60 nm range from 620 to 680 nm

3.1.1.3 Gradient Index Lenses

Gradient index lenses are also used to collimate or focus laser diode beam, particularly to couple laser diode beams into single mode fibers. There are two types of gradient index lens. One is radial gradient index lens, as shown in Fig. 3.3a. This type of gradient index lens can focus a laser beam with two planar surfaces. Another type is axial gradient index lens, as shown in Fig. 3.3b. This type of gradient index lens needs a curved front surface to initiate focusing. The advantage

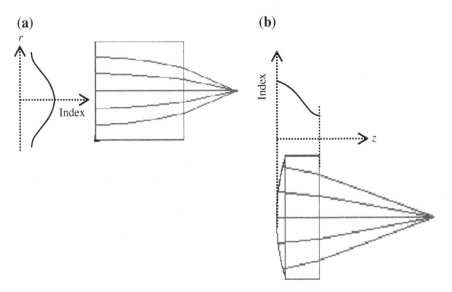

Fig. 3.3 a A Go!FOTON SELFOC® Grin lens focuses a beam. **b** A LightPath Gradium lens focuses a beam

of gradient index lens is that they can perform the same focusing task using milder curvature surfaces with less spherical aberration. However, gradient index lenses with spherical surface do not perform as well as aspheric lenses, and cost less.

3.1.2 Beam Shape Evolvement

The waist of a single TE laser diode beam is elliptical with the major axis of the ellipse in the slow axis direction as shown in Fig. 3.4a. The beam diverts faster in the fast axis direction than in the slow axis direction, because the beam divergence is inversely proportional to its waist size. As the beam propagates, the beam shape becomes circular at certain distance, as shown in Fig. 3.4b. This certain distance depends on the beam waist size and wavelength, and is usually several microns. Beyond this distance, the beam continues diverting faster in the fast axis direction and the beam shape gradually transforms to elliptical again with the major axis in the fast axis direction, as shown in Fig. 3.4c.

If the beam of a single TE mode laser diode is collimated, the major axis of the waist of the collimated beam is in the fast axis direction, as shown in Fig. 3.4d. Then the divergence of the collimated beam is smaller in the fast axis direction than in the slow axis direction. As the collimated beam propagates, the beam shape becomes circular at certain distance, as shown in Fig. 3.4e. This certain distance depends on the collimated beam waist size and the wavelength, and is usually in the range of meters to tens of meters. Beyond this certain distance, the beam shape will become elliptical again with the major axis in the slow axis direction, as shown in Fig. 3.4f.

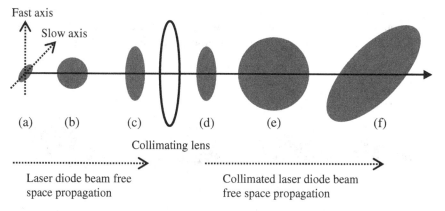

Fig. 3.4 Shape of a single TE mode laser diode beam evolves as the beam propagates in free space and through a collimating lens

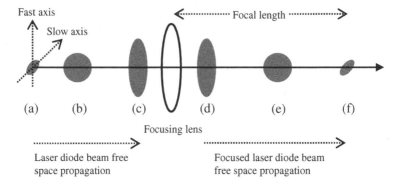

Fig. 3.5 Shape of a laser diode beam evolves as the beam propagates in free space and through a focusing lens

If the beam of a single TE mode laser diode is focused, the focused spot is the image of the beam waist. The major axis of the focused spot is in the slow axis direction, as shown in Fig. 3.5f. Somewhere in-between the focusing lens and the focused spot, the beam shape is circular, as shown in Fig. 3.5e.

3.1.3 Beam Quality Check

Both the lens quality and the alignment accuracy of the laser diode beam to the lens will affect the quality of a collimated or focused beam. The quality of a collimated single TE mode laser diode beam can be checked by visually observing the far-field beam pattern tens of meters away from the laser diode. If a laser diode beam is well aligned to a high-quality aspheric lens with a large NA of 0.5 or up, the collimated beam should have a clean symmetric spot with few very weak diffraction rings and little scattering as shown in Fig. 3.6a. If the laser diode beam is well aligned with a

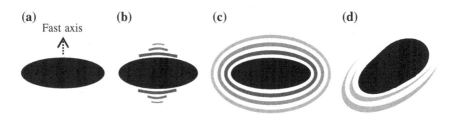

Fig. 3.6 Five far-field spot patterns of collimated laser diode beams. **a** The laser diode beam is well collimated by a good quality aspheric lens with a large NA. **b** The laser diode beam is well collimated by a good quality aspheric lens with a small NA. **c** The laser diode and the lens are transversely well aligned, but longitudinally not well aligned. **d** The laser diode is not well aligned both transversely and longitudinally

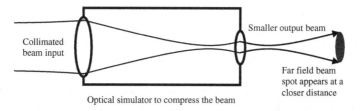

Fig. 3.7 Schematic of an optical simulator that brings far field of a beam closer

high quality aspheric lens with small NA of 0.3 or lower, there will be diffraction rings mainly in the fast axis direction as shown in Fig. 3.6b, because the lens truncates the beam in the fast axis direction. Figure 3.6c shows a symmetric beam spot with several diffraction rings around it. Such a spot pattern indicates that the laser diode is well aligned transversely, but not well positioned longitudinally to the lens, as there are severe spherical aberrations that create interference fringes. Figure 3.6d shows an asymmetric spot with asymmetric diffraction rings around it. Such a spot pattern indicates that the laser diode is not well aligned both transversely and longitudinally. Some low-quality aspheric lenses have less smooth lens surface and will create visually noticeable scattering around the beam spot. The quality of a focused spot of a single TE mode laser diode beam can be checked at the focal plane of the focusing lens using the same criterion shown in Fig. 3.6, because the focal plane is optically at the far field.

An optical simulator can be used to bring the far field of a collimated beam to one meter distance or even closer. The simulator is a beam compressor or a beam expander used in a reversed way, as shown in Fig. 3.7. As we can see from Eq. (3.3) the Rayleigh range of a beam is proportional to the square of the beam waist radius, the compressed beam has a smaller beam waist, a shorter Rayleigh range, and a shorter far-field distance. With a simulator, the setup for observing beam quality can be much more compact.

3.1.4 Collimation or Focusing a Laser Diode Beam, Graphical Explanations

In Sect. 2.3.3, we discussed the analytical mathematical model; thin lens equation; for treating the collimation or focusing of laser beams. In this section we draw several graphs to provide a direct and clear view of the collimating or focusing of a laser diode beam.

3.1.4.1 Collimation

We start from the collimating situation. Figure 3.8a shows an input laser beam with its waist located at the focal plane of the lens, the beam is collimated after propagating through the lens, and the waist of the collimated beam is located at the other focal plane of the lens. Figure 3.8b shows the waist of the input laser beam moves away from the focal plane of the lens by a small distance, the waist of the beam propagated through the lens also moves away from the lens focal plane, this beam is not well collimated. Figure 3.8c shows the waist of the input laser beam moves away from the lens to a location with $o = f + z_R$, then the waist of the beam propagated through the lens reaches the maximum distance i_{max}. Figure 3.8d shows the waist of the input laser beam moves further away from the lens, the waist of the beam propagated through the lens starts moving back toward the lens. We note that the waist size of the beam output from the lens changes in Fig. 3.8a–d, and that the waist positions of the beam output from the lens are the same as in Fig. 3.8b, d, but the waist sizes are different.

When the waist of the input laser beam moves from the lens focal plane toward the lens by a small distance, the beam propagated through the lens is still divergent with the imaginary beam waist appearing on the left-hand side of the lens, as shown in Fig. 3.8e. Figure 3.8f shows the waist of the input laser beam moving closer to the lens to a location with $o = f - z_R$, then the imaginary waist of the beam propagated through the lens reaches the minimum focusing distance i_{min}. Figure 3.8g shows the waist of the input laser beam moving further closer to the lens, the imaginary waist of the beam propagated through the lens starts moving back toward the lens. Figure 3.8h shows the waist of the input laser diode beam is located at $(o/f - 1)^2 = 0.5$ with $z_R^2/f^2 = 0.5$. According to Eqs. (3.5) and (3.6), we have $i = o$ and $w_0' = w_0$, respectively, the beam is relayed with a 1:1 magnification. In Fig. 3.8a–h, the value of z_R/f is smaller than 1, the situation can be categorized as collimation situation. It is noted that the drawings in Fig. 3.8 are only for illustration purpose, they do not have exact proportions.

3.1.4.2 Focusing

The focusing characteristics of a laser diode beam are illustrated in Fig. 3.9. We start from the same situation as shown in Fig. 3.9a, where the waist of the input laser beam is located at the focal plane of the lens, but the value of z_R/f is larger than 1, that is, focusing. The beam propagated through the lens is focused with its waist located at the other focal plane of the lens. Figure 3.9b shows the waist of the input laser beam moves away from the lens to a location with $o = f + z_R$, the waist of the beam propagated through the lens reaches the maximum focusing distance i_{max}. Figure 3.9c shows the waist of the input laser beam moves further away from the lens, the waist of the beam propagated through the lens starts moving back toward the lens.

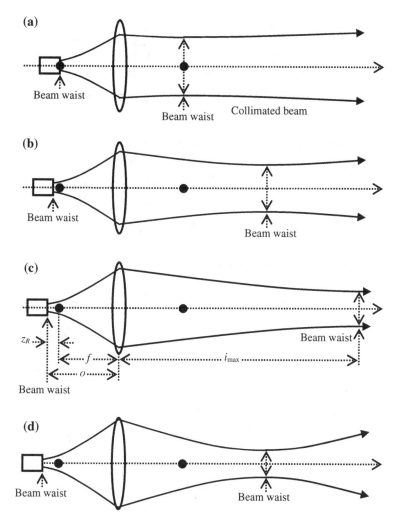

Fig. 3.8 Illustration of the collimating characteristics of a laser diode beam. The *black dots* denote the focal points of the lens

Figure 3.9d shows the input laser beam with its imaginary waist on the right-hand side of the lens as shown by the dot curves (this means if there is no lens, the laser beam waist will reach there), the waist of the beam propagated through the lens also moves away from the lens focal point toward the lens. Figure 3.9e shows the imaginary waist of the input laser beam moving toward the lens from the right-hand side to a position with $o = f - z_R$, the waist of the beam propagated through the lens reaches the minimum focusing distance i_{min}, the negative sign of o indicates that the waist of the input laser beam is on the right-hand side of the lens.

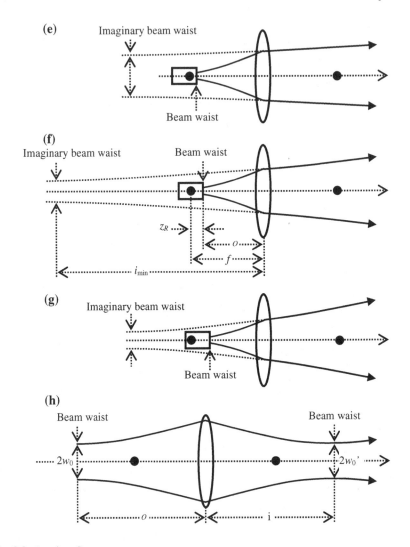

Fig. 3.8 (continued)

Figure 3.9f shows the waist of the input laser beam moving further toward the lens, the waist of the beam propagated through the lens starts moving away from the lens back toward the lens focal point.

We note that in Fig. 3.9a–f, $z_R/f > 1$, the situation can be categorized as focusing situation. The drawings in Fig. 3.9 are only for illustration purpose, they do not have exact proportions.

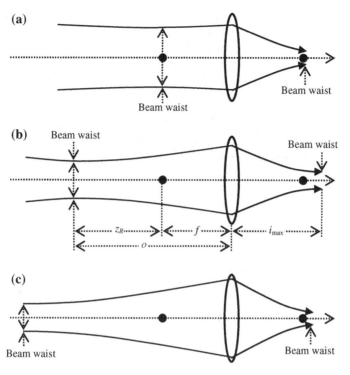

Fig. 3.9 Illustration of the focusing characteristics of a laser diode beam. The *black dots* denote the focal points of the lens

3.1.5 A Numerical Example of Collimating and Focusing a Laser Diode Beam

A numerical example can help to better understand the collimation and focusing characteristics of laser diode beams. We consider a 0.67 μm laser diode beam with FWHM divergence of 24° and 9° in the fast and slow axes directions, respectively. These two angles can be converted into $1/e^2$ intensity full divergence of 40.8° and 15.3°, respectively, by multiplying a coefficient of 1.7. The beam waist $1/e^2$ radii in the fast and slow axis directions can be found using Eq. (3.4) to be $w_{0F} = 0.6$ μm and $w_{0S} = 1.6$ μm, respectively. We assume $M^2 = 1$ and the beam astigmatism $A = 0$ for simplicity. This laser diode beam is collimated by a lens with a focal length of $f_1 = 10$ mm and then focused by another lens with a focal length $f_2 = 20$ mm, the two lenses are apart by a distance of $f_1 + f_2 = 30$ mm, as shown in Fig. 3.10.

From Eq. (3.3) we find the Rayleigh ranges in the fast and slow axis directions to be $z_{ROF} \approx 1.7$ μm and $z_{ROS} \approx 12$ μm, respectively. We can use Eq. (3.6) to calculate the waist sizes, in the fast and slow axis directions, of the beam collimated by lens 1 with $o_1 = f_1$. The results are $w_{1F} \approx 3.53$ mm and $w_{1S} \approx 1.33$ mm. We also have $i_1 = f_1$ from Eq. (3.5), the waist of the collimated beam is at the focal plane of lens 1 and is also

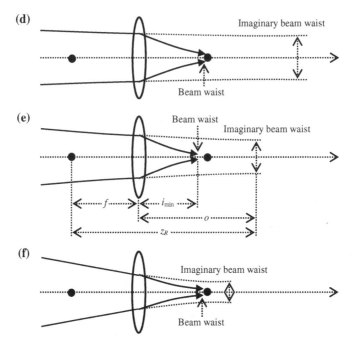

Fig. 3.9   (continued)

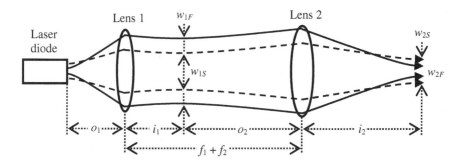

Fig. 3.10   A laser diode beam is collimated by lens 1 and then focused by lens 2. The *solid curves* and *dash curves* are for the beams in the fast and slow axis directions, respectively

located at the focal plane of lens 2 for this setup. The Rayleigh ranges of the collimated beam can be found using Eq. (3.3) to be $z_{R1F} \approx 58.4$ m and $z_{R1S} \approx 8.3$ m, respectively. Finally, the waist sizes, in the fast and slow axis directions, of the beam focused by lens 2 can be found using Eq. (3.6) with $o_2 = f_2$. The results are $w_{2F} \approx 1.2$ μm and $w_{2S} \approx 3.2$ μm. We can see that w_{2F} and w_{2S} are twice larger than w_{0F} and w_{0S} because the setup has a transverse magnification of $f_2/f_1 = 2$. Equations (3.5) and (3.6) for lens 2 are plotted in Fig. 3.11a, b for further analysis.

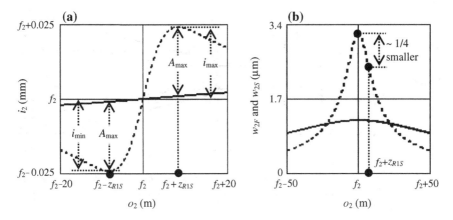

Fig. 3.11 A collimated laser diode beam is focused by a lens with 20 mm focal length. **a** i_2 versus o_2 curve. **b** w_{2F} and w_{2S} versus o_2 curves. In both **a** and **b** the *solid* and *dash curves* are for the fast and slow axis directions, respectively. A_{max} is the maximum "ghost astigmatism"

From Fig. 3.11a we can see an interesting phenomenon; if $o_2 \neq f_2 = 20$ mm, the waists of the focused beam in the fast and slow axes can appear at different positions (different i_2 values for the solid and dash curves), although the laser diode considered in this case has no astigmatism. We call this phenomenon "ghost astigmatism." If the spacing between lens 1 and lens 2 is not $f_1 + f_2$, but much larger, and $o_2 = f_2 \pm z_{R1S}$, we will find $A_{max} \approx 24$ μm using Eq. (3.5) or by directly reading off Fig. 3.11a. Consider the setup shown in Fig. 3.10 has a longitudinal magnification of $(f_2/f_1)^2 = 4$, $A_{max} \approx 24$ μm will lead to an erroneous astigmatism of 24 μm/4 = 6 μm inside the laser diode, that is, a large error.

From Fig. 3.11b we can see another interesting phenomenon; when $o_2 = f_2$, the beam is best focused with the focused spot sizes being the largest in both the fast and slow axis directions. If we increase o_2 from the focusing position of $o = f_2$ to $o_2 = f_2 + z_{R1S}$, the position of the focused spot in the slow axis direction will move from $i_2 = f_2$ to $i_{max} \approx f_2 + 0.024$ mm, and the focused spot is about one-fourth smaller as shown in Fig. 3.11b. That sounds weird, but it is the characteristics of Gaussian beams.

3.1.6 Deliver a Smallest Beam Spot to a Long Distance

Laser diode users often want to deliver a laser beam to a certain distance with a smallest possible spot for highest power density and want to know how small the spot can be. They often use the term "to collimate the beam" or "to focus the beam at this distance" to describe such a requirement. In some cases one of these two terms is correct. In some other cases, neither a collimated beam nor a focused beam can deliver the smallest possible spot to a certain distance. Figure 3.12 shows three

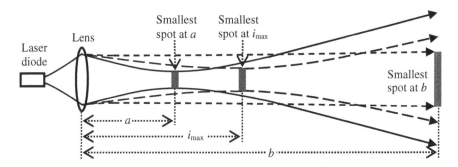

Fig. 3.12 At a long distance b, the collimated beam shown by the *short dash curves* offers the smallest spot. The other two beams are focused at different distances and offer the smallest spots at shorter distances

examples, where the solid curve is for a beam focused at a short distance a, the long dash curve is for a beam focused at the maximum focusing distance i_{max}, and the short dash curve is for a collimated beam. For a long delivery distance, usually beyond tens of meters, a collimated beam can provide the smallest spot, as shown in Fig. 3.12 by distance b. Then, what will be the smallest spot if the delivery distance is larger, but not much larger than i_{max}? In this section we perform an analysis. The result found is simple and easy to use. We only analyze the situation in one transverse direction, either in the fast axis or in the slow axis direction. The situation in another transverse direction can be analyzed the same way.

If we vary the distance o between the laser diode and the lens, as shown in Fig. 3.13, the waist radius $w_0(o)$ and waist location $i(o)$ of the beam output from the lens vary, and the beam radius $w(o)$ at a delivery distance d vary too. Since $w_0(o)$ and $i(o)$ are functions of o, $w(o)$ is also a function of o with d being a parameter. $w(o)$ can be calculated using by

$$w(o) = w_0(o) \left\{ 1 + \left[\frac{M^2 \lambda (d - i(o))}{\pi w_0(o)^2} \right]^2 \right\}^{0.5} \tag{3.8}$$

Equation (3.8) is obtained by replacing z by $d - i(o)$ in Eq. (3.1), as shown in Fig. 3.13.

Assume a laser diode with $\lambda = 0.635$ μm, $M^2 = 1$ and $1/e^2$ intensity full divergence of $2\theta = 23°$, and a focusing lens with $f = 10$ mm. From Eq. (3.6) we find $w_0 = 1$ μm, from Eq. (3.7) we find $i_{max} = 10.12$ m. Combining Eqs. (3.5), (3.6) and (3.8), we plot $w(o) \sim o/f$ by solid curves in Fig. 3.14 for $d = 5$, 10, 20, 50 and 100 m, respectively. The $i(o) \sim o/f$ curve of Eq. (3.5) is also plotted by a dash curve in Fig. 3.14 for the convenience of analysis. The smallest value points on the $w(o) \sim o/f$ curves for all five d values are marked by solid circles. The corresponding points on the $i(o) \sim o/f$ curve for the same o values are marked by open circles. Now, we analyze Fig. 3.14.

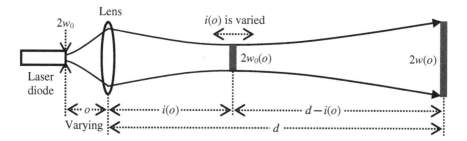

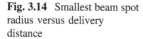

Fig. 3.13 Moving the laser diode along the optical axis to find the smallest spot size $2w(o)$ of the beam at a certain delivery distance d

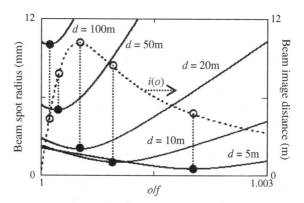

Fig. 3.14 Smallest beam spot radius versus delivery distance

For $d = 5$ m, we can find that the smallest spot radius is 0.5 mm and the corresponding $i(o)$ value is ~ 4.8 m, as shown in Fig. 3.14. This result means that to deliver a beam to 5 m (about half of the maximum focusing distance of 10.12 m) with the smallest spot size, the beam needs be focused at ~ 4.8 m < 5 m.

For $d = 10$ m, we find that the smallest spot radius is 1 mm and the corresponding $i(o)$ value is slightly beyond 8 m. This result means that to deliver a beam to 10 m (close to the maximum focusing distance) with the smallest spot size, the beam needs be focused at ~ 8 m < 10 m.

For $d = 20$ m, the smallest spot radius is 2 mm and the corresponding value on the $i(o) \sim o/f$ is the maximum focusing distance of 10.12 m. This result means that to deliver a beam to 20 m (about twice the maximum focusing distance) with the smallest spot size, the beam needs be focused at the maximum focusing distance 10.12 m $\ll 20$ m.

For $d = 50$ m, the smallest spot radius is 5 mm and the corresponding value on the $i(o)$ value is about 7 m. This result means that to deliver a beam to 50 m (about five times of the maximum focusing distance) with the smallest spot size, the beam needs be near collimated, not focused, with the beam waist located at 7 m away from the lens.

For $d = 100$ m, the smallest spot radius is 10 mm and the corresponding value on the $i(o) \sim o/f$ curve is about 4 m. This result means that to deliver a beam to 100 m (about

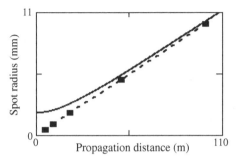

Fig. 3.15 The *solid curve* is the radius of a laser diode beam collimated by a certain lens as a function of propagation distance. The *dash line* is the asymptote of the collimated beam. The smallest spot radius this laser diode combined with this lens can offer is given by the *solid line* for any beam delivery distance

10 times of the maximum focusing distance) with the smallest spot size, the beam still needs be near collimated with the beam waist located at 4 m away from the lens.

The five smallest spot radii shown in Fig. 3.14 are plotted in Fig. 3.15 by open squares versus the beam delivery distance d. If the beam is collimated, the waist radius can be calculated using Eq. (3.6) to be 2.02 mm. The propagation characteristics of such a collimated beam can be found using Eq. (3.1) and is also plotted by dash curve in Fig. 3.15 for comparison. From Fig. 3.15 we can see that the five data points are along a straight line and the straight line is the asymptote of the collimated beam radius. This result means that given a laser diode and a collimating/focusing lens, the smallest possible spot size at any distance can be easily found by drawing an asymptote line to the collimated beam. This result is very interesting and can significantly simplify the process of designing laser beam delivery optics.

3.1.7 Laser Line Generator

3.1.7.1 Macro Lines

1. Using a glass rod or cylindrical lens

Laser light lines are used in many constructions, manufacturing, and medical alignment applications. Laser diode is often the choice of the laser to generate such a laser line. The simplest way to generate a laser line is to collimate the beam first, then use a cylindrical lens or a glass rod to divert the laser beam in one direction, as shown in Fig. 3.16. The laser line fan angle ϕ as well as the line length b is proportional to the collimated beam size a. The elliptical shape beams of laser diodes offer an easy adjustment of line length by rotating the beam about its optical axis. Laser line generators are usually specified by fan angle.

Fig. 3.16 Generate a laser line with fan angle ϕ using a cylindrical lens or a glass rod

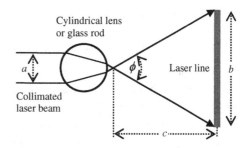

Applications using laser lines are illumination types, not imaging types. The light intensity distribution along the line is the main concern. When the intensity profile of the collimated laser beam is Gaussian, the intensity distribution of the laser line generated by a glass rod or a cylindrical lens is near Gaussian with slightly longer tails. Many applications require a more uniform intensity distribution.

2. Using a Powell prism

In order to generate a laser line with more uniform intensity distribution, a special Powell prism was invented as shown in Fig. 3.17a. With an aspheric cylindrical tip, the Powell prism has more power at the central part and can more strongly spread the stronger central part of the laser beam. As a result the laser line generated has more even intensity distribution. Figure 3.15b shows a Zemax-simulated intensity distribution along the laser line generated by the prism shown in Fig. 3.17a, where 1 million rays are launched along the laser line and 100 sampling pixels are used to plot the curve. The Powell prism is made of H-ZF72 glass of CDGM brand, which has a conic surface with 1 mm radius and -3 conic parameter. The input beam is a Gaussian beam with 2 mm $1/e^2$ intensity size, the laser line generated has FWHM fan angle of about 62°.

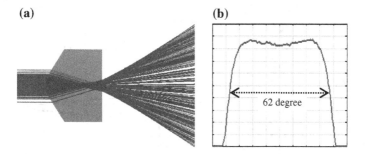

Fig. 3.17 **a** Raytracing diagram of a Powell prism generates a laser line with intensity distribution more even than a Gaussian beam. **b** A Zemax-simulated intensity distribution along the laser line generated by the prism shown in (**a**)

3.1.7.2 Micro Lines

1. Using two aspheric lenses and one weak cylindrical lens

In applications such as flow cytometry, a micro laser line of several microns length is required. Such a micro line can be obtained by increasing the astigmatism in a focused beam. This can be realized by using a weak cylindrical lens in a focused beam as shown in Fig. 3.18 [1].

2. Using one aspheric lens and two cylindrical lenses

Since most off-shelve laser diode lenses are designed to collimate the beam, it is sometime more convenient to use two orthogonal cylindrical lenses in a collimated beam to obtain micro lines, as shown in Fig. 3.19 [1].

3.2 Numerical Analysis of the Propagation, Collimation, and Focusing of a Laser Diode Beam

In Sect. 2.3.4 we discussed the criterion of determining whether a beam is paraxial or nonparaxial. A beam with $1/e^2$ intensity full divergence of 21° starts showing nonparaxial characteristics and a beam with $1/e^2$ intensity full divergence $> 42°$ behaves differently from a paraxial Gaussian beam. The analytical math tools we discussed so far in this book; the thin lens equation and the paraxial Gaussian model; are no longer accurate to treat nonparaxial Gaussian beams. Also, non-paraxial beams have larger divergence and are more likely being truncated by a lens. The truncation will cause complex diffraction that cannot be treated by

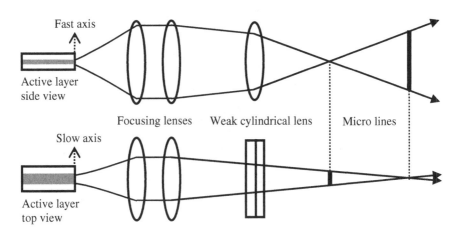

Fig. 3.18 Schematic of a setup uses two aspheric lenses to focus a beam, then uses one cylindrical lens to generate two micro lines

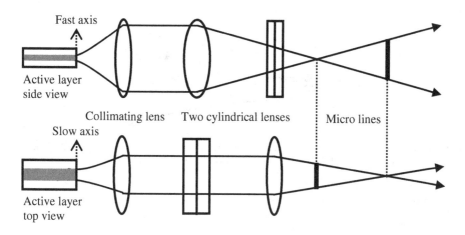

Fig. 3.19 Using two orthogonal positioned cylindrical lenses to generate two micro lines

analytical analysis. If we want accurate results, we have to perform numerical analysis using the Kirchhoff diffraction equation. In this section, we numerically analyze the propagation, collimation, and focusing of laser diode beams.

3.2.1 Propagation of a Nonparaxial Laser Diode Beam

Since laser diode beams can be nonparaxial only in the fast axis direction, we only need to consider a two-dimensional situation.

Consider a laser diode with wavelength $\lambda = 0.635$ μm and a beam waist radius $w_0 = 0.3$ μm in the fast axis direction. The two-dimensional Kirchhoff equation looks like that [2]

$$U(y,z) \sim \left| \int_{-a}^{a} A(y') \frac{e^{-\frac{2\pi i}{\lambda}s}}{s} \left(\frac{z}{s} + 1 \right) dy' \right|^2 \qquad (3.9)$$

where $U(y, z)$ is the intensity of the diffracted field at an observing point (y, z) as shown in Fig. 3.20, $A(y)$ is the amplitude of the source field at $z = 0$, s is the distance between a point y' on the source and an observing point (y, z), and a is the integration range that must cover all the source area. We have

$$s = [(y - y')^2 + z^2]^{0.5} \qquad (3.10)$$

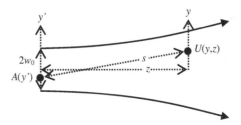

Fig. 3.20 Illustration of a two-dimensional diffraction case. The beam waist is located at the y' plane. The observing point is at the y plane. The two planes are apart by a distance z

In the case considered here, we also have

$$A(y') = e^{-\frac{y'^2}{w_0^2}} \qquad (3.11)$$

Inserting Eqs. (3.10) and (3.11) into Eq. (3.9) and randomly selecting a large enough $a = 1$ μm $> 3w_0$, we perform the integration using MathCAD and obtain the accurate intensity profile $U(y, z)$. Plotted by the solid curve in Fig. 3.21 is the normalized intensity profile $U(y, 5$ mm$)/U(0, 5$ mm$)$ at a randomly selected distance $z = 5$ mm.

Using the paraxial Gaussian model Eq. (3.1), we can find for this beam the $1/e^2$ radius $w(z)$ and the Gaussian intensity profile $e[-2y^2/w(z)^2]$ at any distance z. The dash curve in Fig. 3.21 shows such an intensity profile at 5 mm distance for comparison. We can see that the central portion of the two curves are not very different, the paraxial Gaussian model intensity profile is slightly broader with a FWHM of 3.96 mm; 4.4 % larger than the nonparaxial FWHM of 3.8 mm obtained by evaluating $U(y, 5$ mm$)/U(0, 5$ mm$)$. The paraxial Gaussian model intensity profile has obvious shorter tails with a $1/e^2$ intensity diameter of 6.74 mm; 9.2 % smaller than the nonparaxial $1/e^2$ intensity radius of 7.42 mm obtained by evaluating $U(y, 5$ mm$)/U(0, 5$ mm$)$. These results show that the FWHM beam size is less

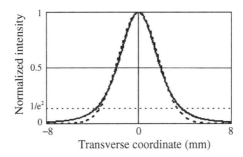

Fig. 3.21 Normalized intensity profiles at 5 mm for a nonparaxial Gaussian beam with a $1/e^2$ intensity radius of 0.3 μm and a wavelength of 0.635 μm. *Solid curve* is obtained using Kirchhoff diffraction integration. *Dash curve* is obtained using paraxial Gaussian model

affected by the beam shape, that is, probably the reason why laser diode industry prefers to use FWHM to define the divergence of the beams. The FWHM and $1/e^2$ far-field divergences of the nonparaxial beam can be found to be 3.8 mm/ 5 mm = 43.5° and 7.42 mm/5 mm = 85°, respectively. This kind of far-field divergences is inside the nonparaxial range defined in Sect. 2.3.4 and is larger than the average divergence in the fast axis direction of laser diodes.

The results of this section indicate that there are differences between the accurate results obtained using Kirchhoff integration and the approximate results obtained using thin lens equation and paraxial Gaussian model. But the differences are not significant. For most applications, we can neglect the nonparaxial nature of laser diode beams. It is noted that the actual intensity profile of the beam waist at the emission facet of a laser diode is not necessary Gaussian. The situation will be more complex. The practical way to analyze such a beam is:

1. Measure the far-field intensity profile of the unmanipulated beam.
2. Try to find an analytical expression to simulate this far-field intensity profile.
3. Use the analytical expression and the Kirchhoff diffraction integration to find the propagation or collimated or focused characteristics of the beam.

3.2.2 Collimating a Nonparaxial Laser Diode Beam

Now we continue to analyze the characteristics of the beam discussed in above section after it is collimated. Assume this beam is collimated by a lens with 5 mm focal length. We already found in above section the intensity profile, $U(y, 5\ mm)/U(0, 5\ mm)$, of this beam at 5 mm distance. The collimating lens converts the beam wavefront from divergent to plane. By principle, we can replace $A(y')$ in Eq. (3.9) by $[U(y, 5\ mm)/U(0, 5\ mm)]^{0.5}$, then perform the numerical integration over the lens area to find the profile of the collimated beam at any distance. But the numerical calculation process may be slow because $U(y, 5\ mm)/U(0, 5\ mm)$ is not an analytical function. Here we try to find an analytical function $F(y)$ to simulate $U(y, 5\ mm)/U(0, 5\ mm)$ in order to speed up the calculation process.

After several times trials, we find $F(y) = \exp[-2y^2/2.89^2 + 0.0254|y^3|]$. The solid and dash curves in Fig. 3.22 show $U(y, 5\ mm)/U(0, 5\ mm)$ and $F(y)$, respectively. The FWHM and $1/e^2$ size of $U(y, 5\ mm)/U(0, 5\ mm)$ and $F(y)$ are the same up to three digits. From Fig. 3.22 we can see that these two curves have no noticeable differences within ±4 mm. While the NA of most real collimation lenses is <0.8, that is, <4 mm radius for a focal length of 5 mm. So, we do not care the behavior of $F(y)$ beyond ±4 mm.

Assuming the collimation lens has a large NA = 0.8, plugging $F(y)^{0.5}$ into Eq. (3.9) and performing integration from −4 to 4 mm, we find the intensity profiles $U'(y, z)$ of the collimated beam at z. For this beam with 3.71 mm $1/e^2$ intensity radius, the Rayleigh range is $\pi w_0^2/\lambda = \pi \times (3.71\ mm)^2/0.635\ \mu m \approx 68$ m, we choose $z = 2$, 10 and 100 m, respectively. The solid curves in Fig. 3.23 show the three

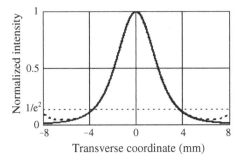

Fig. 3.22 The *solid curve* is $U(y, 5 \text{ mm})/U(0, 5 \text{ mm})$ and the *dash curve* is a simulation function $F(y)$. These two curves have no noticeable differences within $\sim |4 \text{ mm}|$

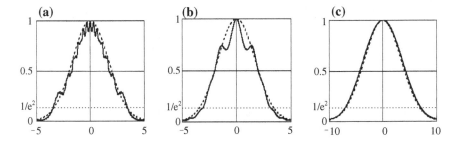

Fig. 3.23 Intensity profiles of a collimated nonparaxial beam at a distance of **a** 2 m, **b** 10 m and **c** 100 m, respectively. *Solid curves* are accurate results obtained using Kirchhoff diffraction integration. The *dash curves* are approximate results obtained by using analytical paraxial Gaussian model. The horizontal axis is transverse variable with a unit of mm. The vertical axis is normalized intensity

normalized intensity profiles $U'(y, 2 \text{ m})/U'(0, 2 \text{ m})$, $U'(y, 10 \text{ m})/U'(0, 10 \text{ m})$ and $U'(y, 100 \text{ m})/U'(0, 100 \text{ m})$, respectively.

In above section, we also use paraxial Gaussian model to obtain a $1/e^2$ intensity radius of 3.37 mm for the beam at 5 mm. Continuing applying paraxial Gaussian model of Eq. (3.1) to this beam, we can find the intensity profiles of the collimated beam at 2, 10 and 100 m, as shown by the dash curves in Fig. 3.23 for comparison purpose. From Fig. 3.23 we can see that both the FWHM and $1/e^2$ intensity sizes of the solid and dash curves are close to within a few percents. In the near field of $z = 2$ and 10 m, the solid curves show ripples that are caused by lens truncation. Analytical thin lens equation cannot predict such a phenomenon.

The results of this section show again that for most applications, the nonparaxial nature of laser diode beams in fast axis direction can be neglected. If we do need to know the details of the intensity profiles, we need to perform numerical analysis using Kirchhoff diffraction integration Eq. (3.9). A detailed analysis on aperture truncation effects will be conducted in Sect. 3.5.

3.2.3 Tightly Focusing an Astigmatic Laser Diode Beam to a Small Spot

In the applications, such as data storage, particle detecting, etc., the beam of a laser diode must be tightly focused to a spot as small as possible. In such a situation, the focused beam is no longer paraxial and the astigmatism in the beam will affect the spot size. Furthermore, most focusing lenses have a NA < 0.6 and will more or less truncate the beam in the fast axis direction. The truncation will cause power loss, increase the focused spot size, shift the focal position [3], and change the beam intensity profile in the fast axis direction. Analytical math tool of thin lens equation cannot effectively handle this type of problem, we must numerically analyze the problem using the diffraction theory [4].

3.2.3.1 Mathematical Model

Figure 3.24a, b illustrates the side view and top view of a fast focusing situation, respectively. The linear polarization of laser diodes is in the slow axis direction and does not affect the focusing in the fast axis direction. Mansuripur [5] showed that when a laser beam is focused with a NA $\leq$ 0.3, the polarization of the beam can be neglected. The NA of laser diode beams in the slow axis direction is usually ~ 0.15. Therefore, the polarization can be completely neglected when tightly focusing a laser diode beam, we can use scalar diffraction theory to study the problem.

The intensity profile of a laser diode beam is elliptical Gaussian at any plane perpendicular to the beam axis. At the lens plane, the elliptical intensity profile can be described by

$$I(x', y') = \exp\left(-2\frac{x'^2}{w_S^2} - 2\frac{y'^2}{w_F^2}\right) \tag{3.12}$$

where w_S and w_F are the $1/e^2$ intensity radii of the beam in the slow and fast axis directions, respectively. If the focusing is slow, we can use Eq. (3.12) to approximate the intensity distribution on the slow converging spherical wavefront. For the fast focusing case considered here, we have to map Eq. (3.12) onto the fast converging spherical wavefront. As illustrated in Fig. 3.25, after some triangle geometry calculation, we find $w_F' = w_F(R_F - z')/R_F$ and

$$a' = \frac{aR_F}{\left(R_F^2 + a^2\right)^{0.5}} \tag{3.13}$$

Similarly, in the x'-z plane, we find $w_S' = w_S(R_S - z')/R_S$. With the results, the elliptical intensity profile of the beam on the converging wavefront takes the form of

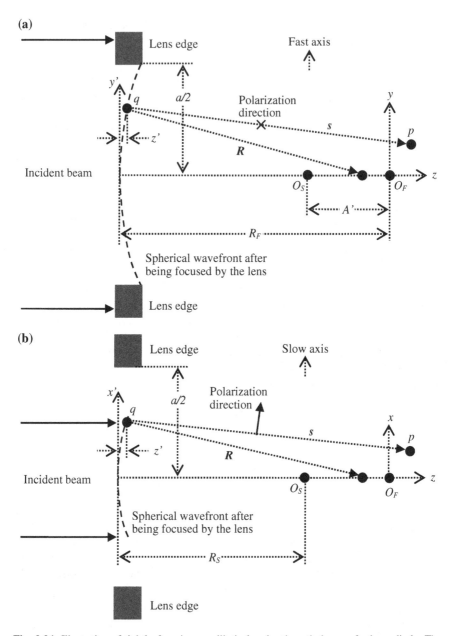

Fig. 3.24 Illustration of tightly focusing an elliptical and astigmatic beam of a laser diode. The beam is linearly polarized in the slow axis direction. O_S and O_F are the wavefront curvature centers of the focused beam in the x'-z and y'-z planes, respectively. R_S and R_F are the wavefront radii of the focused beam in the x'-z and y'-z planes, respectively. A' is the astigmatism of the focused beam. The diameter of the focusing lens aperture is a. The beam is truncated in the fast axis direction. s is the vector linking pint q on the wavefront and point p in the image space. $\boldsymbol{R}$ is the radius vector of point q

Fig. 3.25 Mapping the
Gaussian intensity profile at a
plane onto a converging
spherical wavefront. The *solid
circles* mark the lens
truncation point and the $1/e^2$
intensity radius on the plane,
respectively. The *open circles*
mark the corresponding points
on the converging spherical
wavefront, respectively

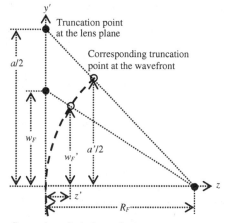

Converging spherical wavefront

$$I(x',y',z') = \exp\left(-2\frac{x'^2 R_S^2}{w_S^2(R_S - z')^2} - 2\frac{y'^2 R_F^2}{w_F^2(R_F - z')^2}\right) \tag{3.14}$$

$$U(x,y,z) \sim \left|\int_S I(x',y')^{0.5}\frac{e^{\frac{2\pi i s}{\lambda}}}{s}[\cos(\phi_s) + \cos(\phi_R)]dS\right|^2 \tag{3.15}$$

When a wave with wavelength λ is incident on an opaque screen with an open
aperture S, the wave is diffracted, the field intensity at a point $p(x, y, z)$ beyond the
screen is described by the well-known Kirchhoff diffraction integral [2], where
$I(x', y')^{0.5}$ is the field amplitude at the lens plane, s is the distance between point
q on the lens plane and point p, ϕ_s and ϕ_R are the angles between vectors $\boldsymbol{n}$ and $\boldsymbol{s}$,
and between vectors $\boldsymbol{n}$ and $\boldsymbol{R}$, respectively, $\boldsymbol{n}$ is the normal vector of the lens plane,
and the two-dimensional integral covers the lens aperture S.

A careful look at Eq. (3.15) reveals that a small error, say <1 %, in the calculation of s in the term of $1/s$ as well as in the terms of $\cos(\phi_s)$ and $\cos(\phi_R)$ will
significantly affect the results. However, since $s \gg \lambda$, an error of <1 % in the
calculation of s can be $>\lambda$, can dramatically change the phase in the term of exp
$(2\pi i s/\lambda)$, and invalid the results. This observation provides the guideline for how to
take approximations.

s is given by

$$s = \left[(R_F - z' + z)^2 + (x' - x)^2 + (y' - y)^2\right]^{0.5} \tag{3.16}$$

$\cos(\phi_s)$ and $\cos(\phi_R)$ are given, respectively, by

$$\cos(\phi_s) = \frac{R_F + z - z'}{s} \tag{3.17}$$

$$\cos(\phi_R) = \frac{(R^2 - x'^2 - y'^2)^{0.5}}{R} \tag{3.18}$$

$$\approx \left(1 - \frac{x'^2 + y'^2}{R_F{}^2}\right)^{0.5}$$

Since we know that $R_S \le R \le R_F$, and $|R_S - R_F| \ll R_S \approx R_F$, we can use R_F to replace R in Eq. (3.18) to simplify the situation. R is of a few millimeters, $|R_S - R_F|$ is of several microns, the error in Eq. (3.18) is <1 %, this is the only approximation taken in this section.

It is noted that the origin of coordinate x-y-z is chosen to be at O_F, and the origin of coordinate x'-y'-z is chosen to be at the vertex of the converging astigmatic wavefront at the lens aperture. In such a coordinator system, the astigmatic (spherocylindrical) wavefront can be described by [6]

$$z' = \frac{x'^2/R_S + y'^2/R_F}{1 + \left[1 - \frac{(x'^2/R_S + y'^2/R_F)^2}{x'^2 + y'^2}\right]^{0.5}} \tag{3.19}$$

The effect of astigmatic wavefront is included by inserting Eq. (3.19) into Eq. (3.16).

In our case, the general Kirchhoff diffraction integral Eq. (3.15) takes the form

$$U(x, y, z) \sim \left| \int_{-\frac{a'}{2}}^{\frac{a'}{2}} \int_{-\left(\frac{a'^2}{4}-y'^2\right)^{0.5}}^{\left(\frac{a'^2}{4}-y'^2\right)^{0.5}} |I(x', y', z')|^{0.5} \frac{e^{\frac{2\pi i s}{\lambda}}}{s} [\cos(\phi_s) + \cos(\phi_R)] dx' dy' \right|^2 \tag{3.20}$$

The integration covers the converging wavefront with radius of $(x'^2 + y'^2)^{0.5} = a'/2$. Combining Eqs. (3.13), (3.14), (3.16)–(3.20), we can numerically solve for $U(x, y, z)$ using a computing software such as MathCAD.

3.2.3.2 Two Design Examples

Consider a focusing lens or lens group that has an $f = 2.5$ mm and is optimized for an application of magnification = 1, we have $o = i$ and $1/o + 1/i = 1/2.5$ mm that leads to $o = i = 5$ mm. Consider using this lens to focus the beam of a Sanyo DL3038-033 laser diode. The datasheet says this laser diode has a wavelength of $\lambda = 635$ nm, 5 mW output power, an astigmatism of 8 μm, and 8° and 35° FWHM divergent angles in the slow and fast axis directions, respectively. This divergence can be converted by a factor of 1.7 to $1/e^2$ intensity full divergence of $2\theta_S = 13.6°$ and $2\theta_F = 59.5°$, respectively. The $1/e^2$ intensity beam radii on the principal plane of the lens in the slow and fast axis directions, respectively, can be calculated by $w_S = 5$ mm × tan$(\theta_S) = 0.60$ mm and $w_F = 5$ mm × tan$(\theta_F) = 2.86$ mm. The $1/e^2$ intensity beam waist radius at the laser diode emission facet in the slow axis direction can be backcalculated by paraxial Gaussian model to be $w_{S0} = \lambda/(\theta_S\pi) = 1.70$ μm, assuming $M^2 = 1$. Because of the large divergence θ_F, the $1/e^2$ intensity beam waist radius at the laser diode emission facet in the fast axis direction can no longer be calculated by paraxial Gaussian model as being discussed in Sect. 3.2.1. We do this calculation later.

The other numbers we can obtain now are $A' = 8$ μm because the lens magnification is 1, $R_F = 5$ mm and $R_S = R_F - A' = 4.992$ mm. Figure 3.26 plots the $1/e^2$ intensity profiles of the beam at the lens in the $x'-z$ plane (slow axis) and $y'-z$ plane (fast axis), respectively, and a weak truncation by a lens with aperture $a = 6$ mm (NA = $0.5a/R_F = 0.6$, 3.7 % energy is truncated) and a strong truncation by a lens with aperture $a = 3$ mm (NA = $0.5a/R_F = 0.3$, 30.5 % energy is truncated). We can see from Fig. 3.26 that a lens aperture with $a = 3$ mm severely truncates the beam in the fast axis direction, but does not truncate the beam in the slow axis direction.

Inserting the values of λ, w_S, w_F, R_S, and R_F into Eqs. (3.13), (3.14), (3.16)–(3.20) and letting $a = 6$ and 3 mm, respectively, and numerically solving these equations using MathCAD, we have obtained $U_{a=6mm}(x, y, z)$ and $U_{a=3mm}(x, y, z)$. An observation reveals that $U_{a=6mm}(0, 0, z)$ has the maximum value on axis at $z = 0$, that means there is no noticeable focal shift in the fast axis direction, and that $U_{a=3mm}(0, 0, z)$ has the maximum value on axis at $z = -0.7$ μm. This means the lens truncation shifts the focal position in the fast axis direction by $= -0.7$ μm. Since $|-0.7$ μm$| \ll |A'| = 8$ μm, this focal shift is far from enough to correct the astigmatism. Two normalized intensity profiles along the axial direction, $U_{a=6mm}(0, 0, z)/U_{a=6mm}(0, 0, 0)$ and $U_{a=3mm}(0, 0, z)/U_{a=3mm}(0, 0, -0.7$ μm$)$, are plotted in Fig. 3.27.

The transverse intensity profiles in slow and fast axis directions for five z values are plotted in Fig. 3.28a–h. $z = 0$ is at the geometrical focal point in the fast axis direction, $z = -0.7$ μm is at the real focal point in the fast axis direction, $z = -8$ μm is at the geometrical as well as the real focal point in the slow axis direction, and

Fig. 3.26 Two apertures with radius $a = 3$ mm and $a = 6$ mm, respectively, truncate the Gaussian beam

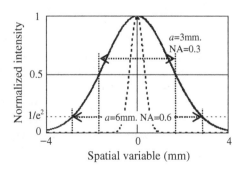

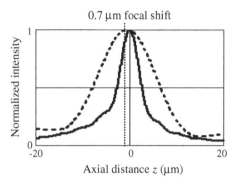

Fig. 3.27 Normalized intensity profiles along the z axis for $a = 6$ and 3 mm, respectively. *Dash curve* $U_{a=6mm}(0, 0, z)/U_{a=6mm}(0, 0, 0)$. *Solid curve* $U_{a=3mm}(0, 0, z)/U_{a=3mm}(0, 0, -0.7 \mu m)$, shows 0.7 μm focal shift in the fast axis direction caused by lens truncation

$z = 20$ μm and $z = 50$ μm are randomly picked to show the intensity profiles beyond the focal region. As we can see in Fig. 3.28 that the intensity profiles in the fast axis direction are complex because of the lens truncation. We further plot the intensity profiles in the fast axis direction for three larger z values in Fig. 3.29. The intensity distribution in any other plane can be plotted by letting $x = py$, where p is a coefficient to be selected.

All the beam intensity profiles in the slow axis direction are shown by the dash curves in Fig. 3.28 are Gaussian, since the beam is not truncated in this direction. The focused spot appears at $z = A' = -8$ μm because of the astigmatism. The $1/e^2$ intensity beam radii shown by the dash curves in Figs. 3.28c and 3.24d are 1.7 μm as expected, matches $w_{S0} = 1.7$ μm for a magnification of 1. The situation is simple.

All the beam intensity profiles in the fast axis are shown by the solid curves in Figs. 3.28 and 3.29. A way from the focal point, these curves start showing ripples caused by the beam truncated in this direction. This phenomenon is more striking for stronger truncation of $a = 3$ mm. The curves at the left-hand side column of Fig. 3.28 is for $a = 6$ mm. The $1/e^2$ intensity radius of the focused spot in the fast axis direction shown by the solid curve in Fig. 3.28a is 0.559 μm.

In order to find the effect of lens truncation effect, we increase the integration circle size in Eq. (3.20) to $a = 9.8$ mm, the beam is no longer truncated in the fast axis direction, the $1/e^2$ intensity radius of the focused spot in the fast axis direction shown by the solid curve in Fig. 3.28a reduces from 0.559 to 0.514 μm, a difference of $(0.559 \mu m - 0.514 \mu m)/0.514 \mu m \approx 9$ %, which is caused by the truncation of the $a = 6$ mm lens aperture. Since the case studied here is magnification = 1, we can say that the $1/e^2$ intensity radius in the fast axis direction at the emission facet of the laser diode is $w_{F0} \approx 0.51$ μm. If we use the paraxial Gaussian mode, we have $w_{F0} = \lambda/(\pi\theta_F) \approx 0.39$ μm, that is, an error of $(0.51 \mu m - 0.39 \mu m)/0.51 \mu m \approx 24$ %.

Away from the focal plane, the intensity profiles shown by the solid curves are not Gaussian, but may still be approximated by Gaussian functions depending on the accuracy desired.

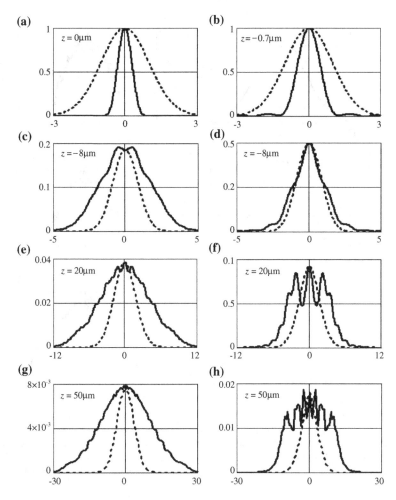

Fig. 3.28 The *solid curves* and the *dash curves* are the normalized intensity profiles in the fast and slow axis directions, respectively, for $z = 0, -0.7, -8, 20$ and $50\ \mu m$. The horizontal axes have unit of microns. The *left* and *right* hand side columns are for $a = 6$ mm and $a = 3$ mm, respectively. The details for each graph are: **a** $U_{a=6mm}(x, 0, 0)/U_{a=6mm}(0, 0, 0)$ and $U_{a=6mm}(0, y, 0)/U_{a=6mm}(0, 0, 0)$, **b** $U_{a=3mm}(x, 0, -0.7\ \mu m)/U_{a=3mm}(0, 0, -0.7\ \mu m)$ and $U_{a=3mm}(0, y, -0.7\ \mu m)/U_{a=3mm}(0, 0, -0.7\ \mu m)$, **c** $U_{a=6mm}(x, 0, -8\ \mu m)/U_{a=6mm}(0, 0, 0)$ and $U_{a=6mm}(0, y, -8\ \mu m)/U_{a=6mm}(0, 0, 0)$, **d** $U_{a=3mm}(x, 0, -8\ \mu m)/U_{a=3mm}(0, 0, -0.7\ \mu m)$ and $U_{a=3mm}(0, y, -8\ \mu m)/U_{a=3mm}(0, 0, -0.7\ \mu m)$, **e** $U_{a=6mm}(x, 0, 20\ \mu m)/U_{a=6mm}(0, 0, 0)$ and $U_{a=6mm}(0, y, 20\ \mu m)/U_{a=6mm}(0, 0, 0)$, **f** $U_{a=3mm}(x, 0, 20\ \mu m)/U_{a=3mm}(0, 0, -0.7\ \mu m)$ and $U_{a=3mm}(0, y, 20\ \mu m)/U_{a=3mm}(0, 0, -0.7\ \mu m)$, **g** $U_{a=6mm}(x, 0, 50\ \mu m)/U_{a=6mm}(0, 0, 0)$ and $U_{a=6mm}(0, y, 50\ \mu m)/U_{a=6mm}(0, 0, 0)$, and **h** $U_{a=3mm}(x, 0, 50\ \mu m)/U_{a=3mm}(0, 0, -0.7\ \mu m)$ and $U_{a=3mm}(0, y, 50\ \mu m)/U_{a=3mm}(0, 0, -0.7\ \mu m)$

The curves at the right-hand side column of Fig. 3.28 is for $a = 3$ mm. The $1/e^2$ intensity radius of the focused spot in the fast axis direction at the focal plane $z = -0.7\ \mu m$ shown by the solid curve in Fig. 3.29b is 0.86 μm, that is, 67 % larger

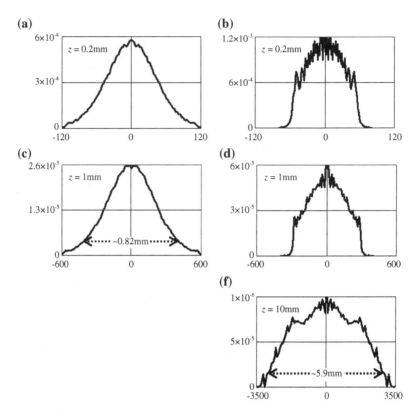

Fig. 3.29 Normalized intensity profiles in the fast axis direction for $z = 200$ μm, 1 mm and 10 mm, respectively. The *left* and *right* hand side columns are for $a = 6$ mm and $a = 3$ mm, respectively. The details for each graph are: **a** $U_{a=6mm}(0, y, 200$ μm$)/U_{a=6mm}(0, 0, -0.7)$, **b** $U_{a=3mm}(0, y, 200$ μm$)/U_{a=3mm}(0, 0, -0.7$ μm$)$, **c** $U_{a=6mm}(0, y, 1$ mm$)/U_{a=6mm}(0, 0, -0.7)$, **d** $U_{a=3mm}(0, y, 1$ mm$)/U_{a=3mm}(0, 0, -0.7$ μm$)$, **e** $U_{a=6mm}(0, y, 10$ mm$)/U_{a=3mm}(0, 0, -0.7$ μm$)$, and **f** $U_{a=3mm}(0, y, 10$ mm$)/U_{a=3mm}(0, 0, -0.7$ μm$)$

than 0.514 μm and is caused by the stronger lens truncation. Even at the focal plane, there are noticeable side lobes, but the central lobe is still near Gaussian. As the axial distance increases, the beam intensity profiles in the fast axis direction shown by the solid curves start to have strong ripples and can no longer be approximated by Gaussian functions. The periods and the amplitudes of the ripples in the intensity profiles gradually decrease as z increases, as shown in Fig. 3.29. However, the beam profiles never regain Gaussian pattern.

The propagation characteristics of the focused beams in the fast axis direction are illustrated in Fig. 3.30, stronger lens truncation causes larger focused spot and smaller far-field divergence. The M^2 factor of the focused beams in the fast axis direction can be estimated by modifying Eq. (3.4)

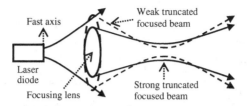

Fig. 3.30 Illustration of the propagation characteristics of the focused beam in the fast axis direction. *Dash curve* is for weak lens truncation. *Solid curve* is for strong lens truncation

$$M^2 = \frac{\theta_F \pi w_{1F}}{\lambda} \qquad (3.21)$$

where w_{1F} is the $1/e^2$ intensity radius of the focused spot and θ_F is the $1/e^2$ intensity far field half-divergence of the beam in the fast axis direction. The $1/e^2$ intensity radius in Fig. 3.29c is estimated to be 0.41 mm, we have $\theta_F \approx 0.41$ mm/ $z = 0.41$ mm/1 mm = 0.41, $w_{1F} \approx 0.56$ μm, and obtain from Eq. (3.21) $M^2 \approx 1.14$. The $1/e^2$ intensity radius in Fig. 3.29f is estimated to be 2.95 mm, we have $\theta_F \approx 2.95$ mm/$z = 2.95$ mm/10 mm = 0.295, $w_{1F} \approx 0.86$ μm, and obtain $M^2 \approx 1.26$. We note that these values of M^2 factor well match our measurement results.

3.3 Beam Circularization and Astigmatism Correction

In some applications, a circular shape laser diode beam is more desirable than an elliptical shape beam. Circularizing laser diode elliptical beams becomes a special technical issue. On the other hand, an astigmatism-free laser diode beam is sometimes beneficial too. Since some widely used techniques can both circularize an elliptical beam and correct the astigmatism, we discuss these two issues in the same section.

3.3.1 Using Two Cylindrical Lenses to Collimate and Circularize an Elliptical Beam and to Correct the Astigmatism

When thinking about collimating, circularizing, and correcting astigmatism for a laser diode beam, the first idea usually coming up is to use a pair of orthogonally positioned cylindrical lenses. As shown in Fig. 3.31, a cylindrical lens with higher power is placed a few millimeters away from the laser diode to collimate the laser diode beam in the fast axis direction, another cylindrical lens with lower power is placed about 10 mm away from the laser diode to collimate the beam in the slow

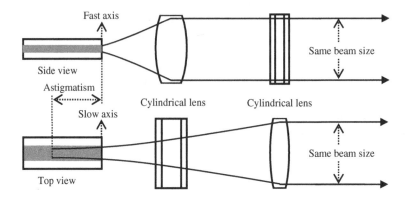

Fig. 3.31 Using two orthogonal positioned cylindrical lenses to collimate and circularize an elliptical laser diode beam as well as to correct the astigmatism of the beam

axis direction. The focal lengths of the two cylindrical lenses are so chosen that after collimation, the beam sizes in the fast and slow axis directions are the same, as shown in Fig. 3.31. Then the beam is collimated, the elliptical beam shape is circularized, and the astigmatic is corrected.

Although the idea shown in Fig. 3.31 sounds great, it has big problems in reality. First, the large divergent angle of the beam in the fast axis direction requires the cylindrical lens to be aspheric. Aspheric cylindrical lenses are difficult to fabricate, rarely seen, and expensive. Second, the higher power cylindrical lens will be thick, and introduce aberration into the beam in the slow axis direction. This effect combined with the not very small divergence of the beam in the slow axis direction also requires the lower power cylindrical lens to be aspheric. Therefore, in reality only multi-TE mode laser diode beams are collimated by a pair of spherical cylindrical lenses, because these high-power multi-TE mode beams are often used for illumination types of applications, where aberration is not a concern.

3.3.2 Using an Anamorphic Prism Pair and a Circular Aperture to Circularize the Beam

As shown in Fig. 3.32, a pair of anamorphic prisms can either expand or compress a laser diode beam in one direction. A collimated elliptical beam of a laser diode can be circularized by either expand the beam in the minor axis direction or compress the beam in the major axis direction. The expansion or compression ratio can be adjusted from about 2–6 by rotating the prisms. After propagating through the prism pair, the beam will have several millimeters transverse displacement as shown in Fig. 3.32. We can see from Fig. 3.32 that one prism can also expand or

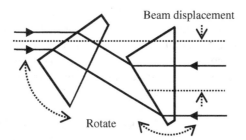

Fig. 3.32 A pair of anamorphic prisms can expand or compress the beam in one direction. Propagating from *left* to *right*, the beam is expanded, and vice versa. The expansion or compression ration can be adjusted by rotating the prisms

compress the beam, only with less expansion or compression ratio and the beam propagation direction will be changed. After being expanded or compressed by anamorphic prism pairs, the beam needs be truncated by a circular aperture.

During the circularizing process, the beam shape and intensity profile will be going through the following changes, as shown in Fig. 3.33:

(a) A laser diode beam has elliptical intensity contours and elliptical shape.
(b) A collimating lens truncates the beam in the fast axis direction.
(c) The collimated beam still has elliptical intensity contours, but not real elliptical shape because of the lens truncation.

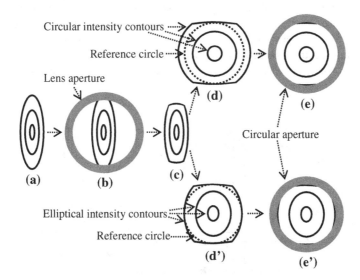

Fig. 3.33 The details of using an anamorphic prism pair and a circular aperture to circularize a collimated laser diode beam. **a–e** or **e'** show step by step how the beam shape and intensity profiles change

(d) An anamorphic prism pair can expand the collimated beam to such a beam that has circular intensity contours, but the beam shape is not real circular.

(e) A circular aperture can cut the beam shown in (d) to circular shape as well as circular intensity contours with relatively more power loss.

(f) The beam shown in (c) can also be expanded by an anamorphic prism pair to such a beam that has near circular shape, but still has elliptical intensity contours.

(g) A circular aperture can cut the beam shown in (f) to circular shape with relatively less power loss, but the intensity contours is still a little elliptical.

A circularize aperture truncating the beam will generate diffraction rings in the far field.

3.3.3 Using a Weak Cylindrical Lens to Correct the Astigmatism

After being collimated by an aspheric lens and circularized by a pair of anamorphic prisms, the astigmatism is still in the beam. Figure 3.34 shows an example. In Fig. 3.34a the beam is still slightly divergent in the fast axis direction and Fig. 3.34b the beam is collimated in slow axis direction. The astigmatism can be corrected by using a weak cylindrical lens with a few meters focal length to collimate the beam in the fast axis direction. Although a negative cylindrical lens can also be used to divert the beam in the slow axis direction. Different types of laser diodes have different astigmatism magnitudes. For a given laser diode, after being collimated by different lenses, the astigmatism in the beam is different. The chance is low to have a weak cylindrical lens with focal length just right to correct the astigmatism. The way of solving this problem is explained in Fig. 3.34c. The optical power of a cylindrical lens can be represented by a vector O that can be decomposed to two

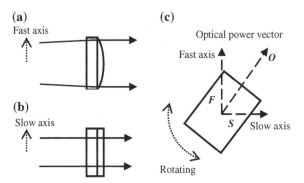

Fig. 3.34 Using a weak cylindrical lens to correct the astigmatism by collimating the beam in the fast axis direction. **a** Side view. **b** Top view. **c** Front view

vectors F and S in the fast and slow axis directions, respectively. The net power of the cylindrical lens for correcting the astigmatism is given by $|F| - |S|$. Rotating the cylindrical lens about the beam axis, the magnitude of $|F| - |S|$ can be continuously varied from $-|O|$ to $|O|$. Using a cylindrical lens with a $|O|$ more than needed, we can always find an angle where the cylindrical lens makes the right correction of the astigmatism. Since the cylindrical lens is very weak, it is difficult to visually evaluate how well the astigmatism is corrected. Some type of wavefront sensing equipment should be used to measure the quality of the final beam.

3.3.4 Using a Special Cylindrical Micro Lens to Circularize an Elliptical Beam and to Correct the Astigmatism

A patented cylindrical micro lens was invented to circularize the elliptical beams of laser diodes and to correct the astigmatism [7]. The micro lens has a diameter of about 100 μm and is placed tens of microns away from the laser diode in the slow axis direction, as shown in Fig. 3.35. The first surface of the lens collimates the beam in the fast axis direction and the second surface of the lens slightly diverts the beam in the fast axis direction. When the prescription of the two surfaces, the thickness and the position of the micro lens, are correct, the beam can be circularized and astigmatism free, as shown in Fig. 3.35. Since the micro lens needs be positioned close to the laser diode, any tiny defects on the lens will be magnified and generate scattering. The beam is also multiple reflected between the two surfaces of the lens because of the small size of the lens. The final beam will look like

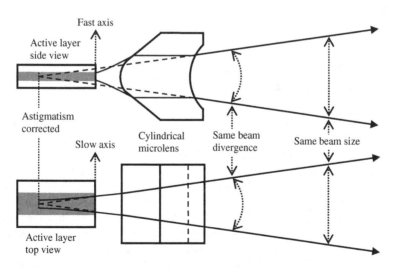

Fig. 3.35 Using a special cylindrical micro lens to circularize the beam and correct the astigmatism

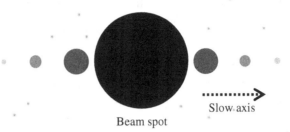

Fig. 3.36 The beam of a Circulaser™ is visually not clean with multiple reflection spots and
scattering around the main spot. However, the beam quality is still good

that shown in Fig. 3.36 and is visually "dirty." Since human eye response to light
intensity is nonlinear, the multiple reflection spots and scattering noticeable to
human eyes are actually very weak. As we mentioned in previous section, laser
diodes of different types have different beam divergence and astigmatism. Even
different laser diodes of the same type can have different beam divergence and
astigmatism. Several slightly different micro lenses are available for choosing from.
The lens position and orientation need to be carefully adjusted, while the beam
quality is real-time monitored by a wavefront sensor.

The final beam has a circularity of 1.2:1 or less. The quality of the circularized
beam in terms of wavefront error is still good, usually with a peak to valley error of
$<\lambda/4$. The micro lens is AR coated, but will still cause about 20 % power loss to the
beam, because the incident angle range of the beam is large, no AR coating can
perform well in such a large angle range. This micro lens is not for retail sale in the
market, it is licensed to Coherent Inc. and Blue Sky Research Inc. to be installed in
their laser diode modules called Circulaser™.

3.3.5 Using a Single Mode Optic Fiber to Circularize
an Elliptical Beam and Correct the Astigmatism

Another way of circularizing the elliptical beam of a laser diode and correcting the
astigmatism is to couple the beam into a single mode optic fiber, as shown in
Fig. 3.37. The beam inside the single mode optic fiber will be transformed to the TE
mode of the fiber. Since the fiber is circular, the beam output from the fiber is also
circular and astigmatism free. The beam output from a single mode fiber is clean,
which has circularity better than 1.1:1 and a wavefront error of about $\lambda/10$. To the
author's experience, the beam output from a single mode fiber has the highest
optical quality compared with the beams of Circulaser™ or the beams circularized
by anamorphic prism pairs. Coupling a laser diode beam into a single mode fiber
with a core size of several microns is not easy, we discuss the technique used in the
section below.

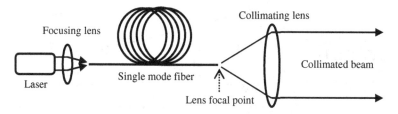

Fig. 3.37 Scheme of a laser beam collimator utilizing single mode fiber

3.4 Coupling a Single Mode Laser Diode Beam into a Single Mode Optic Fiber

Single mode optic fiber can well circularize the elliptical beam of laser diodes, correct the astigmatism, deliver the beam to a location where a laser diode beam cannot directly reach, and therefore is widely used. Various laser diode modules with single mode fiber output are available in market, often with a nickname as "pigtail" modules. However, the coupling of a laser diode beam into a single mode fiber is not very easy.

Single mode fiber has a very small core size, about 6 μm for red color, and about 12 μm for 1.55 μm light. The light field inside a fiber is also called "mode." The fiber core size and the difference between the indexes of the fiber core and the cladding surrounding the core determine the mode size. The output beam of a single mode fiber is very close to a Gaussian beam with the beam waist at the fiber output facet. The beam far-field divergence is inversely proportional to the beam waist size and is about 10° or so FWHM. Single mode fiber core also has an acceptance cone given by [8]

$$\theta = \sin^{-1}\left[\left(n_{\mathrm{co}}^2 - n_{\mathrm{cl}}^2\right)^{0.5}\right] \tag{3.22}$$

where n_{co} and n_{cl} are the refractive indexes of the fiber core and cladding, respectively, as shown in Fig. 3.38. The acceptance cone is often expressed as NA with NA = sin θ. Most single mode optic fiber has an acceptance NA in the range of 0.1–0.15. Some special optic fibers can have acceptance the NA from 0.05 to 0.4. To couple a single TE mode laser beam into a single mode optic fiber without big power loss, a laser diode beam must be focused to a spot smaller than the fiber core size, and at the same time the beam convergent cone must be smaller than the fiber acceptance cone. Unfortunately, most single TE mode laser diode beams cannot meet these two requirements simultaneously. When a beam is focused in the fast axis direction to match the core size and acceptance cone of the fiber, the beam in the slow axis direction will have a focused spot larger than the core, as shown Fig. 3.38. When a beam is focused in the slow axis direction to match the core size and acceptance cone of the optic fiber, the beam in the fast axis direction will have a convergent cone angle larger than the fiber acceptance cone, as shown Fig. 3.38. This phenomenon is a result of the elliptical beam.

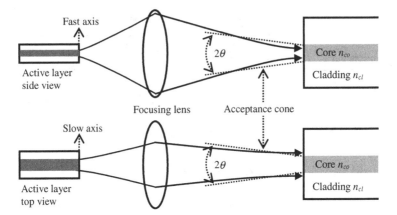

Fig. 3.38 Illustration of coupling a single TE mode laser diode beam into a single mode fiber, where θ is the acceptance cone angle of the fiber. When the beam in the slow axis direction is matched to the fiber core and acceptance cone, the convergent cone of the beam in the fast axis direction will be larger than the fiber acceptance cone. Similarly, when the beam in the fast axis direction is matched to the fiber core and acceptance cone, the focused spot size of the beam in the slow axis direction will be larger than the fiber core

To couple a single TE mode laser diode beam into a single mode fiber without big power loss, the beam must also be well aligned to the fiber core. This involves three linear and two angular adjustments. The three linear adjustments are more critical than two angular adjustments. The two transverse adjustments are more critical than the longitudinal adjustment. A high-quality x-y-z translation stage is required for the alignment, and some patience and experience are required as well. The practical coupling efficiency is about 50 %. The beam from a Circulaser^{TM} is circular and can be coupled into a single mode fiber with efficiency up to about 80 %. However, the Circulaser^{TM} has about 20 % power loss caused by the micro lens to start with. As a comparison, gas laser beams are circular and can be focused to simultaneously match the core and acceptance cone of a single mode fiber, the coupling efficiency can be over 90 %.

It is noted that the divergent angle of a beam output from a single mode fiber is different from the acceptance cone angle of the fiber, since the cone angle given by Eq. (3.22) does not depend on the fiber mode size, while the divergent angle of the beam output from the fiber is inversely proportional to the mode size.

The tip of a single mode fiber can be chemically treated to form a lens, as shown in Fig. 3.39, such a fiber is called butt-end fiber. The fiber tip lens can focus the divergent beam of a laser diode and couple the beam directly into the fiber. The fiber must be placed within a few microns to the laser diode facet, since the beam has a large divergence. Butt-end fiber is not easy to find in market.

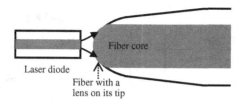

Fig. 3.39 A butt-end single mode optic fiber with a lens on its tip for directly coupling the beam of a single TE mode laser diode into the fiber

3.5 Aperture Beam Truncation Effects

Single TE mode laser diode beams have divergences in the fast axis direction larger than the NAs of most collimating lenses, therefore a beam will often be truncated by the lens edge. Such a truncation on beam will severely change the near-field intensity profile of the beam, create ripples in the profile and increase the far-field divergence of the beam. As the beam propagates, its intensity profile will be gradually changing and eventually changes to near Gaussian. Below we study an example.

Consider a one-dimensional case: a collimated Gaussian beam with $1/e^2$ intensity radius of 2 mm and a wavelength of 0.635 μm is truncated by two apertures with diameters of 2 and 4 mm, respectively. The situation is illustrated in Fig. 3.40. The intensity profile of the beam after being truncated can be calculated by the two-dimensional Kirchhoff diffraction equation, Eq. (3.9), we rewrite it here

$$U(y,z) \sim \left| \int_{-a/2}^{a/2} e^{-\frac{y'^2}{2\,mm^2}} \frac{e^{\frac{2\pi i}{\lambda}\left[(y-y')^2+z^2\right]^{0.5}}}{\left[(y-y')^2+z^2\right]^{0.5}} \left\{ \frac{z}{\left[(y-y')^2+z^2\right]^{0.5}} + 1 \right\} dy' \right|^2 \quad (3.23)$$

where the factor 1 indicates that the input beam is collimated with a flat wavefront, the cosine of the wavefront normal and the lens aperture normal is 1.

Figure 3.41 shows the normalized intensity profile $U(y, z)/U(0, z)$ for six z values and aperture diameter of $a = 2$ mm and $a = 4$ mm, respectively. The $1/e^2$ intensity

Fig. 3.40 A one-dimensional case of an aperture truncates a collimated beam

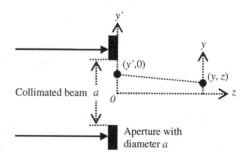

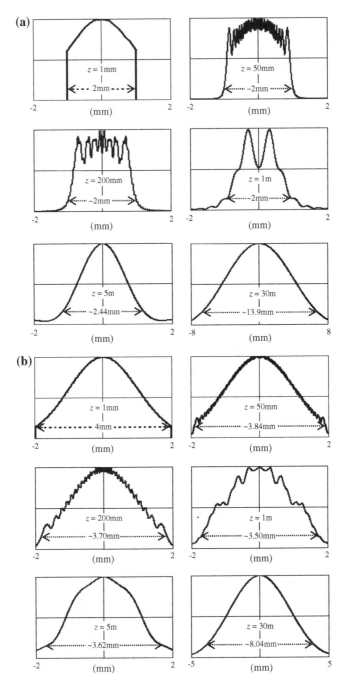

Fig. 3.41 The normalized intensity profile $U(y, z)/U(0, z)$ for $z = 1$, 50, 200 mm, 1, 5 and 30 m, respectively. The horizontal axis is the radial variable. The beam is a collimated beam with $1/e^2$ intensity radius of 2 mm. The truncation aperture diameter is **a** 2 mm and **b** 4 mm

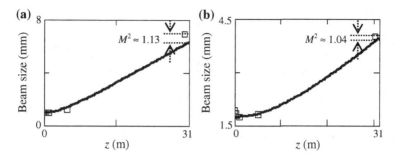

Fig. 3.42 Beam propagation characteristics. The *open square* symbols are for the truncated beams. *Solid curves* are for the untruncated beams with the same waist size for comparison. **a** 2 mm aperture diameter. **b** 4 mm aperture diameter

radii of the truncated beam are also calculated and marked in Fig. 3.41. As we can see that the beam intensity profiles are not Gaussian in the near field, are near Gaussian in the far field, and the smallest $1/e^2$ intensity radii are ~ 1 and ~ 1.75 mm for the 2 and 4 mm diameter apertures, respectively.

We can model the propagation characteristics of the truncated beam by plotting the $1/e^2$ intensity radii versus z, as shown by the open square symbols in Fig. 3.38. The propagation characteristics of two untruncated Gaussian beams with the $1/e^2$ intensity waist radii of 1 and 1.75 mm are also plotted in Fig. 3.42 by the solid curves for comparison. Comparing the far-field size of the truncated and untruncated beam, we find $M^2 = 1.13$ and 1.04 for the 2 and 4 mm aperture diameters, respectively.

3.6 Diffractive Optics and Beam Shaping

3.6.1 Diffractive Optics

Various diffractive optical elements have been developed to generate various beam patterns, to focus the beam like a conventional lens, or to shape a beam. Figure 3.43 shows several beam patterns that are often seen. Diffractive optics can generate much more patterns than those shown in Fig. 3.43. Diffractive optics has a micro structure in it and utilizes diffractions to perform the designed tasks. Diffraction effects take certain distance to work. If two diffractive optics are placed in a row, the second diffractive optics will likely interfere the performance of the first diffractive optics, and the beam will be a mess, such a setup should be avoided. Diffractive optics can work only for certain wavelength, incident angle, and usually for a collimated beam. They are rarely used in a complex optical path with many conventional lenses in it.

Fig. 3.43 Some commonly seen beam patterns generated by diffractive optics

3.6.2 Shape a Single TE Mode Laser Diode Beam

For single TE mode laser diode, beam shaping often means transform a Gaussian beam to a flat-top or top-hat beam, because many applications require a flat-top beam. Figure 3.44a shows a two-lens beam shaper. The first lens is a planar-conic lens with 3 mm radius and −11 conic parameter. The second lens is a biconvex spherical lens with 60 mm radius on both sides. Both lenses are made of BK7 glass. Similar to the Powell prism, the conic surface of the first lens has more power at the central portion and spreads the beam more in the stronger central portion, thereby generates more even intensity distribution. The second lens collimates the beam. The input beam is a 2 mm size Gaussian beam with a wavelength of 0.635 μm. The collimated beam size is about 17 mm. Figure 3.44b, c shows Zemax simulated intensity profiles of the collimated beam at 100 and 200 mm distance, respectively. The sampling pixel number is 100×100, and 10^7 analysis rays are launched. As we can see that the intensity profile varies as the beam propagates. Any beam shaper has a limited working range, usually half meter or so. Beyond this range, the beam intensity profile will gradually change to a quasi-Gaussian profile because of diffraction.

3.7 Effects of External Optical Feedback on Laser Diodes

3.7.1 Accidental External Feedback

The emitted power of a laser diode can be accidently or intentionally fed back into the laser diode active layer. A feedback of as low as 1 % of the emitted power can make the laser diode operation unstable. A higher level feedback can significantly increase the laser output power and possibly blow out the laser diode.

Figure 3.45 shows two examples of accidentally external feedback. In Fig. 3.45a a laser diode beam is focused onto an optical surface, such as a CD surface, with the waist of the focused beam on the surface. Since the beam has no divergence at its waist and the beam is incident on the surface in the normal direction, the reflected beam will trace the incident path back into the active layer of the laser diode. If the feedback is unwanted, the optical surface should be tilted by a few degrees to deflect the reflected beam. Figure 3.45b shows an optical surface being positioned

(a)

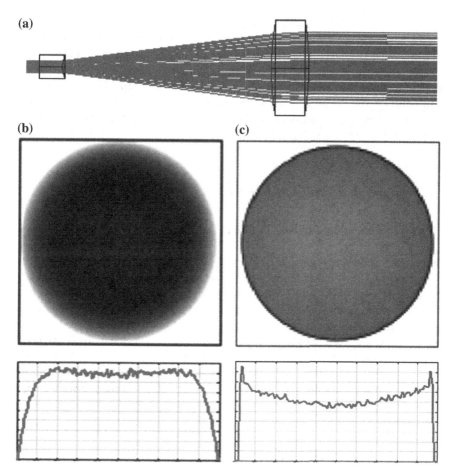

(b) **(c)**

Fig. 3.44 **a** A two-lens flat-top beam shaper. Intensity profiles of the flat-top beam at **b** 100 mm and **c** 200 mm distance

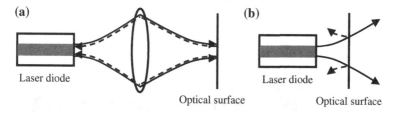

Fig. 3.45 **a** *Solid curves* are a single TE mode laser diode beam focused on an optical surface. *Dash curves* are the beam reflected by the optical surface back into the active layer. **b** *Solid curves* are a single TE mode laser diode beam incident on an optical surface positioned close to the laser diode. *Dash curves* are the beam reflected back by the optical surface

very close to the laser diode output facet. Although the beam reflected by the optical surface has large divergence, a certain portion of the laser power can still be fed back into the active layer. Since in such a situation the beam size is not much larger than the active layer size, the feedback level can be significant, and the laser diode operation will be disturbed. Such a situation should be avoided whenever possible. If an optical surface must be placed very close to a laser diode, some type of laser power monitoring and controlling mechanism should be used to protect the laser diode from possible over driven.

3.7.2 Intentional External Feedback

3.7.2.1 Fiber Grating External Feedback

Figure 3.46 shows an example of intentional external optical feedback, a fiber grating external cavity laser diode system. The laser diode has one facet HR coated and another facet AR coated, as shown in Fig. 3.46. A butt-end fiber with a refractive grating build in its core is positioned near the laser diode. The fiber grating and the HR-coated facet of the laser diode form a lasing cavity. The fiber grating will reflect portion of the laser power incident on it back into the active layer of the laser diode and output another portion of the laser power. The grating has a narrow spectral bandwidth, only the wavelength inside the bandwidth will be reflected, and the laser diode is forced to operate at this wavelength with a very narrower linewidth.

3.7.2.2 Grating External Feedback

Figure 3.47 shows two Littman configuration grating external feedback wavelength tunable laser diode systems. In Fig. 3.47a, the laser diode has its two facets HR and AR coated, respectively. The beam emitted from the laser diode is first collimated by a lens, and then incident on a grating. The zero-order diffraction or the reflected beam of the grating is the output beam. The first order diffraction beam of the

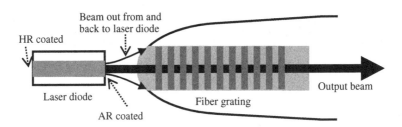

Fig. 3.46 Schematic of a fiber grating external cavity laser diode system

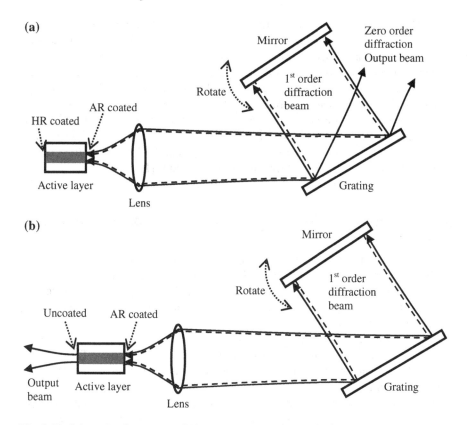

Fig. 3.47 Schematic of two types of Littman grating external cavity laser system

grating is incident on a mirror and the mirror reflects the beam back to the laser diode active layer via the grating and the lens. The mirror and the HR-coated facet of the laser diode form the lasing cavity. Because of the dispersion function of the grating, the propagation direction of the first order diffraction is a function of wavelength. The mirror at certain angle will only reflect certain wavelength back to the laser diode and force the laser to operate at this wavelength. The laser wavelength can be tuned by rotating the mirror.

In Fig. 3.47b, the laser diode has one facet AR coated and another facet is uncoated. The beam emitted from the laser diode AR-coated facet is first collimated by a lens, and then incident on a grating. The first order diffraction beam of the grating is incident on a mirror and the mirror reflects the beam back to the laser diode active layer via the grating and the lens. The mirror and the uncoated facet of the laser diode form the lasing cavity. The laser outputs through the uncoated facet. The laser wavelength can be tuned by rotating the mirror.

There are two great advantages of grating external feedback laser diode system over a simple laser diode:

1 The lasing wavelength can be tuned by tuning the mirror angle. The tuning range is limited by the active medium bandwidth. For 1.55 μm laser diode, the tuning range can be over 100 nm.

2 The linewidth is dramatically reduced. Equation (1.4) in Sect. 1.3.5 tells us that the laser linewidth is inversely proportional to the optical length of the lasing cavity. The lasing cavity formed by the mirror and the HR coated or the uncoated facet of the laser diode can be over 100 mm long, that is, over 100 times longer than the laser diode active layer cavity of about 1 mm optical length. Therefore, the linewidth is dramatically reduced [9]. The narrow bandwidth of the grating can further reduce the linewidth [10].

3.8 Multitransverse Mode Laser Diode Beam Manipulation

To increase the laser power, the active layer width has to be increased. Such laser diodes are wide-stripe laser diodes. A beam of wide-stripe laser diodes contains many TE modes as shown in Fig. 2.2. Each TE mode is a Gaussian beam. In the fast axis direction, there is only one TE mode, the beam is still Gaussian, same as a single TE mode laser beam. We have already discussed the characteristics and manipulation of single TE mode laser diode beam in previous sections. In the slow axis direction, there are many TE modes. The combination of all these TE modes is no longer Gaussian, is more like a geometric line shape light source, and needs be treated in a different way.

To further increase the laser power, several wide stripes can be piled up to form a laser diode stack. There are many TE modes in both fast and slow axis directions. The combination of all these TE modes make behaves like a geometric, rectangular shape light source. To avoid confusion, we call the combination of all the TE modes "beam," the characteristics of the beam is more important to users than the characteristics of individual mode.

3.8.1 Collimating a Wide-Stripe Laser Diode Beam

Multi-TE mode laser diode beams have high power and are mainly used in illumination type of applications where less accurate beam manipulation is required. Yet, multi-TE modes laser diode beams sill need to be collimated or focused before they can be effectively used. The often asked question is how small the overall spot will be at a certain working distance when the beam is collimated? In this section, we discuss this topic in this section.

Figure 3.48 shows a schematic of collimating, in the slow axis direction, a wide-stripe multi-TE mode laser diode beam, where a is the half-size of the wide stripe, w_0 is the radius of the TE mode waist in the slow axis direction or in the wide strip direction, the value of w_0 is usually a couple of microns much smaller than a, s is

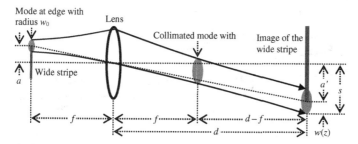

Fig. 3.48 Collimating a multi-TE mode laser diode beam using a lens. The *dash lines* are for geometric optics. The *solid curves* are for the laser mode. The proportion of the drawing is not accurate for illustration purpose

the overall spot size at working distance d and is what we want to find. For simplicity, we assume the mode at the edge of the stripe is aligned with the stripe in a way shown in Fig. 3.48. We will soon see that this assumption does not make any real difference.

When the beam is collimated by a lens, the wide strip as well as the waist of the modes are placed at the focal plane of the lens, a imposes an angle Ω to the optical axis of the lens, as shown in Fig. 3.48. At a working plane with a distance d away, the half-size of the image of the wide stripe is given by

$$\begin{aligned} a' &= d\Omega \\ &= d(a/f) \end{aligned} \tag{3.24}$$

where f is the lens focal length. When the edge mode arrives at the working plane, it is centered at the edge of the image of the wide stripe, as shown in Fig. 3.48.

Now we consider an example. The wide stripe has a half-size of $a = 100$ μm, the working distance is $d = 100$ m and the collimating lens has a focal length of $f = 10$ mm. We have $\Omega = a/f = 100$ μm/10 mm $= 10$ mR, $a' = 10$ mR $\times d = 10$ mR $\times 100$ m $= 1$ m. In Sect. 3.1.6, we calculated the size of a typical laser diode TE mode at various distances, and found the mode radius to be 10 mm at 100 m. Here, we just adopt this result, have $w(z) = 10$ mm $\ll a'$ and $s \approx a'$. This result means that when collimating the beam of a wide-stripe laser diode in the slow axis direction, we can neglect the TE mode and simply treat the beam as a geometric light source. Thereby the calculation can be significantly simplified. It is noted that the beam of a wide-stripe laser diode is still Gaussian in the fast axis direction.

3.8.2 Focusing a Wide-Stripe Laser Diode Beam

We consider a case of using one lens to focus the beam of a wide-stripe laser diode at a working distance of $d = 500$ mm. The schematic of the focusing situation is shown in Fig. 3.49. The wide stripe, the edge mode, and the focusing lens are the

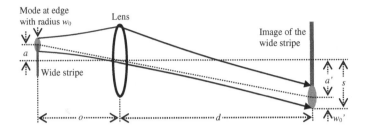

Fig. 3.49 Focusing a multi-TE mode laser diode beam using a lens. The *dash lines* are for geometric optics. The *solid curves* are for the laser mode. The proportion of the drawing is not accurate for illustration purpose

same as those used in Sect. 3.8.1. s is the total spot size and is what we are trying to find. From the geometric optics thin lens equation Eq. (2.8), we find $o = 10.204$ mm will have the wide stripe being focused at 500 mm.

Then, we have from Eq. (3.24) that $a' = d(a/o) = 500$ mm $\times$ (100 μm/ 10.204 mm) = 4.9 mm. Consider a laser diode with TE mode radius in the slow axis direction being $w_0 = 1.0$ μm and wavelength $\lambda = 0.635$ μm, from the thin lens equation for Gaussian beam, Eq. (3.5), we find $o = 10.204$ mm will lead to the TE mode being focused at 499.91–500 mm. From Eq. (3.6) we find that $w_0' = 49$ μm $\ll a' = 49$ mm, we have $s \approx a'$. The conclusions drawn here are the same as the conclusion drawn in Sect. 3.8.1 above, that is, when focusing the beam of a wide-stripe laser diode in the slow axis direction, we can neglect the TE mode and simply treat the beam as a geometric light source.

3.8.3 Collimating or Focusing a Laser Diode Stack Beam

The schematic of a laser diode stack is shown in Fig. 2.4. The stack has multi-TE modes in both fast and slow axis directions. In Sects. 3.8.1 and 3.8.2, we found that when collimated or focused, the TE mode size of a wide-stripe laser diode in the slow axis direction is much smaller than the size of the wide-stripe image and the effects of these TE modes can be neglected. The collimated or focused TE mode size in the fast axis direction is even smaller than the size in the slow axis direction. Therefore, a laser diode stack can be treated as a rectangular shape geometric light source with size $a \times b$ in both the slow and fast axis directions. Equation (3.24) can be applied to both slow and fast axis directions.

3.8.4 Wide-Stripe Laser Diode Beam Shape Evolvement

The divergence of a wide-stripe multi-TE mode laser diode beam is large in the fast axis direction. If the beam starts with a line shape shown in Fig. 3.50a, as the beam

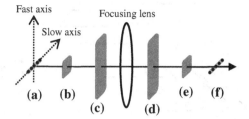

Fig. 3.50 The shape of a wide-stripe laser diode beam evolves as the beam propagates in free space and through a lens. The *tiny black dots* in the wide stripe and image line represent the transverse modes and their images, respectively

propagates, the beam size in the fast axis direction increases faster than the beam size in the slow axis direction. At one certain distance, the beam shape becomes a square with round corners due to diffraction, as shown in Fig. 3.50b. As the beam continues propagating, the beam shape becomes a vertical rectangular with round corners, as shown in Fig. 3.50c. If the beam is focused by a lens, the focused spot is the image of the wide stripe, as shown in Fig. 3.50f, and at certain location in between the lens and the focused image, the beam shape is square as shown in Fig. 3.50e. If the focusing lens is of good quality, we may still be able to see the image of these transverse modes in the image line, as shown by black dots in Fig. 3.50f. But more often, image of all these transverse modes merge together because lens aberrations increase the size of the mode image.

3.8.5 Laser Diode Stack Beam Shape Evolvement

The shape of a laser diode stack beam evolves as the beam propagates in a way similar to the shape of a wide-stripe laser diode beam evolves. Within a few microns to the laser diode stack, the beam has rectangular shape as shown in Fig. 3.51a, either a horizontal or vertical rectangle depends the stack structure. As the beam propagates, its shape gradually transforms to a vertical rectangle with round corners due to diffraction, as shown in Fig. 3.51b. If the beam is collimated or

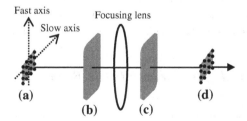

Fig. 3.51 The shape of a laser diode stack beam evolves as the beam propagates. The *tiny black dots* in the laser stack and the image *rectangle* represent the transverse modes and their images, respectively

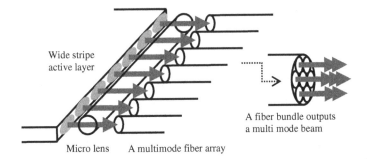

Wide stripe
active layer

A fiber bundle outputs
a multi mode beam

Micro lens   A multimode fiber array

Fig. 3.52 A cylindrical micro lens or glass fiber focuses a multi-TE mode laser diode beam into a multimode fiber array, the fiber array then forms a fiber bundle to output a *circular shape* multi-TE mode beam

focused by a lens, the image of the laser stack will have a shape similar to the stack as shown in Fig. 3.51d. If the lens is of good quality, we may still be able to see the image of these transverse modes shown by black dots in Fig. 3.51d.

3.8.6 Coupling a Multitransverse Mode Laser Diode Beam into a Multimode Optic Fiber Array

A Multi-TE mode wide-stripe laser diode beam can be picked up by a multimode fiber array as shown in Fig. 3.48, where a micro lens (glass fiber) of 100 μm diameter or so is used to focus the modes in the fast axis direction onto the fiber array. Multimode fiber has a large core size, often larger than 50 μm. After the modes propagating through the micro lens, the mode sizes in the slow axis direction can still be smaller than the fiber core. Coupling multi-TE mode laser diode beam into multimode optic fiber is not as difficult as coupling single TE mode laser diode beam into single mode optic fiber, the coupling efficiency can be high. The fiber array is then rearranged to a circular shape fiber bundle to output a circular multi-TE mode beam, as shown in Fig. 3.52, since a circular shape beam is often more desired than a linear shape beam. The beam output from every fiber contains multi-TE modes, the output beam from the fiber bundle contains a large number of TE modes. Such a beam is more like a geometric light source than a laser beam. The output beam of fiber bundle can be collimated or focused by a lens. The collimated or focused spot size can be calculated using the technique described in Sect. 3.8.1.

3.8.7 Shape a Wide-Stripe Laser Diode Beam

The focused spot shape of a wide-stripe laser diode is a light line; that is often not desirable. Various beam shaping optical devices have been invented to optically transform the one-dimensional light source of wide stripe to a two-dimensional light

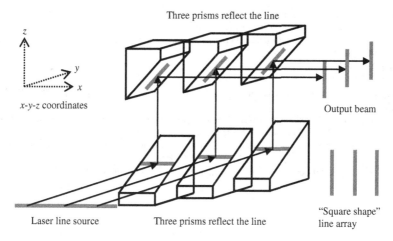

Fig. 3.53 Illustration of a beam shaping prism array. A laser line is broken into three sections and forms a "*square shape*" line array

source. Most such devices are a micro prism array. Figure 3.53 shows an example. A line shape light source propagating in the y direction is incident on an array of three micro prisms. The three prisms are positioned in such a way that they break the light line into three equal length sections and reflect the three light line sections upward in the z direction. Then the three line sections are incident on another array of three micro prisms as shown in Fig. 3.53. The second prism array reflects the three line sections in the x direction as the output beam. Looking in the negative x direction, we will see the three line sections aligned in parallel and form a "square shape" line array. Since the lines have smaller divergence in the line direction and larger divergence in the direction perpendicular to the lines, they will merge together to form a square shape beam after propagating through certain distance.

The beam shaper shown in Fig. 3.53 is an illustration. There are many patented beam shapers. For example, one beam shaper can transform a light line into five sections and form a "square shape" five line array. Compared with the three line array, the five line array has more even spatial intensity distribution in the direction perpendicular to the lines, and are a better light source for many applications. The difficult task is not only to invent a compact beam shaper on paper, but also to fabricate it at reasonable cost.

References

1. Patents US6478452, US6612719, US20020196562, WO2001054239 A1 and EP1252692A1
2. http://en.wikipedia.org/wiki/Kirchhoff's_diffraction_formula
3. Sucha, G., et al.: Focal shift for a Gaussian beam: an experimental study. Appl. Opt. **23**, 4345–4347 (1984)

4. Sun, H.: A simple mathematical model for designing laser diode focusing optics with large numerical aperture. Opt. Eng. **53**, 105105-1–105105-7 (2014)
5. Mansuripur, M.: Vector diffraction theory of focusing in systems of high numerical aperture. Opt. Photon News **3**, 72–75 (1992)
6. Menchaca, C., et al.: Toroidal and spherocylindrical surfaces. Appl. Opt. **25**, 3008–3009 (1986)
7. Snyder, J.J., et al.: Fast diffraction-limited cylindrical microlenses. Appl. Opt. **30**, 2743–2747 (1991)
8. Many optics text books dealing with fiber optics have relevant contents. For example, Saleh, B.E.A., Teich, M.C.: Photonics, Chapter 8 Fiber Optics, Wiley, New York (1991)
9. Sun, H., et al.: Calculation of spectral linewidth reduction of external-cavity strong-feedback semiconductor lasers. Appl. Opt. **33**, 4771–4775 (1994)
10. Loh, H., et al.: Influence of grating parameters on the linewidths of external-cavity diode lasers. Appl. Opt. **45**, 9191–9197 (2006)

Chapter 4
Laser Diode Beam Characterization

Abstract Techniques for characterizing the spatial and spectral properties of single TE mode laser diode beams are described. The spatial properties include beam size and shape, waist size and location, M^2 factor, far field divergence, and astigmatism. The spectral parameters include the wavelength and linewidth. The measurement of laser power and energy is briefly touched.

Keywords Astigmatism · Beam size · Characterization · Divergence · Fizeau wedge · Grating · Interference fringes · Scanning Fabry–Perot interferometer · Laser power · Linewidth · M^2 factor · Waist size · Waist location · Wavelength

Characterizing laser diode beams includes three aspects:

1. Characterizing the spatial properties.
2. Characterizing the spectral properties.
3. Measuring the power/energy.

Characterizing the first two aspects of laser diode beams is more complex than characterizing those of the other types of lasers and are the emphasis of this chapter. Measuring the power or energy of laser diodes is similar to measuring the power or energy of other types of lasers.

4.1 Spatial Property Characterization

Since multi-TE mode laser diode beams cannot be well collimated or focused to a small spot, these beams are mainly used for illuminations, their spatial properties do not need to be characterized to a high accuracy. Single TE mode laser diode beams are often used in applications where precise collimation or tight focusing is required. Therefore, characterizing the spatial properties of single TE mode laser diode beams is important. Different types of laser diodes often have different spatial properties and the laser diodes of the same type can have different spatial properties.

© The Author(s) 2015
H. Sun, *A Practical Guide to Handling Laser Diode Beams*,
SpringerBriefs in Physics, DOI 10.1007/978-94-017-9783-2_4

Therefore, characterizing single TE mode laser diode beams is more complex than characterizing the beams of other types of lasers.

Single TE mode laser diode beams have large divergence and are almost never directly used. These beams are often collimated by an optical system before being used. Therefore characterizing single TE mode laser diode beams often means to characterize the collimated beams. In this section, we mainly discuss the techniques for characterizing the spatial properties of collimated single TE mode laser diode beams. The spatial properties of unmanipulated beams of laser diodes can be back calculated.

Five parameters describe the spatial properties of a single TE mode laser diode beam, they are: beam waist radius in the fast and slow axis directions, respectively, beam waist locations in the fast and slow axis directions, respectively, and the M^2 factor.

4.1.1 Beam Size and Shape Measurement

Laser beam profilers are instruments used to characterize the size and shape of laser beams. There are two types of laser beam profilers: camera based and scanner based. These beam profilers are mainly used for characterizing laser diode beams since laser diode beams can have various different spatial properties.

A camera-based beam profiler consists of a two-dimensional sensing array for catching the beam and a computer loaded with special software for data processing and display. The advantage of using a 2D sensing array is that it can provide a true 2D picture of the beam. Two types of 2D sensor arrays can cover the spectral range from 190 to 1,550 nm. These sensor arrays are easy to find in market and are not expensive. Sensor arrays for wavelength longer than 1,550 nm can be expensive, which is a disadvantage. Another disadvantage of the camera-based beam profiler is the relative low spatial resolution limited by the pixel size of a few microns. The software and computer can display various beam parameters, such as $1/e^2$ intensity diameter, $1/e^2$ intensity encircled power, beam center position, etc.

A scanner-based beam profiler consists of a scanner and a computer loaded with special software for data processing and display. Figure 4.1a shows a widely used knife edge beam scanner. A rotor has a right-angle, triangle-shape knife on it; the beam to be scanned is focused by a lens. The focused beam propagates through the rotor and incidents on a single element sensor. Since the knife has a right angle shape, as the rotor spins, the beam is simultaneously scanned in the two orthogonal directions, as shown in Fig. 4.1b. The beam intensity profiles in these two orthogonal directions can be calculated to a resolution of submicron based on the scanner position and the sensor output signal. Such a spatial resolution is higher than the spatial resolution of pixel-based 2D sensing arrays. It is also less difficult to find different types of single element sensor to cover wide wavelength ranges. The disadvantage of a scanning beam profiler is that the beam intensity profiles in the directions other than the two orthogonal scanning directions cannot be directly

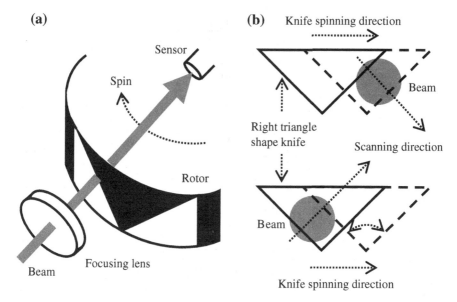

Fig. 4.1 Illustration of a knife edge beam profiler

measured. The computer software will data fit the beam intensity profiles in these other directions and display the best guess results. The computer and software can display various beam parameters, such as $1/e^2$ intensity diameter, $1/e^2$ intensity encircled power, beam center position, etc.

4.1.2 Locate the Beam Waist and Measure the M^2 Factor

In this section, we need use Eqs. (2.1), (2.2), (2.4), (2.5), (2.10), (2.14), (2.27), and (2.28). For the convenience of the readers, we rewrite these eight equations here:

$$w(z) = w_0 \left[1 + \left(\frac{M^2 \lambda z}{\pi w_0^2} \right)^2 \right]^{1/2} \tag{4.1}$$

$$R(z) = z \left[1 + \left(\frac{\pi w_0^2}{M^2 \lambda z} \right)^2 \right] \tag{4.2}$$

$$z_R = \frac{\pi w_0^2}{M^2 \lambda} \tag{4.3}$$

$$\theta = \frac{w(z)}{z}$$

$$= \frac{M^2 \lambda}{\pi w_0}$$

$$= \frac{w_0}{z_R} \tag{4.4}$$

$$\frac{i}{f} = \frac{\frac{o}{f}\left(\frac{o}{f}-1\right)+\left(\frac{z_R}{f}\right)^2}{\left(\frac{o}{f}-1\right)^2+\left(\frac{z_R}{f}\right)^2} \tag{4.5}$$

$$m = \frac{w_0{}'}{w_0}$$

$$= \frac{1}{\left[\left(\frac{o}{f}-1\right)^2+\left(\frac{z_R}{f}\right)^2\right]^{0.5}} \tag{4.6}$$

$$w_0 = \frac{w(z)}{\left[1+\frac{w(z)^4 \pi^2}{R(z)^2 (M^2 \lambda)^2}\right]^{0.5}} \tag{4.7}$$

$$z = \frac{R(z)}{1+\frac{R(z)^2 (M^2 \lambda)^2}{w(z)^4 \pi^2}} \tag{4.8}$$

where $w(z)$ is the $1/e^2$ intensity radius of a Gaussian beam at distance z from the beam waist, w_0 is the $1/e^2$ intensity radius of the beam waist, M^2 is the M square factor, λ is the wavelength, $R(z)$ is the wavefront radius at z, z_R is the Rayleigh range, θ is the far field divergence, f is the focal length of the collimating or focusing lens, o is the object distance defined as the axial distance between the lens principle plane and the waist position of the beam incident on the lens, i is the image distance defined as the axial distance between the lens principle plane and the waist position of the beam collimated or focused by the lens, m is the lens magnification, and $w_0{}'$ is the $1/e^2$ intensity radius of the waist of the collimated or focused beam.

Equations (4.7) and (4.8) indicate that if we have measured at certain axial positions the beam radius $w(z)$ and the beam wavefront radius $R(z)$, and knew the wavelength λ and the M^2 factor, we can back calculate the beam waist radius w_0 and the distance z between the beam waist and the measurement location. Measuring the beam size is relatively simple as described in the above section. Unfortunately, measuring the wavefront radius requires the use of an interferometer that is not available in many labs.

If we have measured the beam radius $w(z)$ at certain axial planes, knew the beam waist radius w_0, the wavelength λ, and the M^2 factor, we can calculate the distance z between the beam waist and the measurement location by modifying Eq. (4.1) to

$$z = \frac{w_0 \pi}{M^2 \lambda} \left[w(z)^2 - w_0{}^2 \right]^{0.5} \tag{4.9}$$

Similarly, if we have measured the beam radius $w(z)$ at certain axial planes, knew the distance z between this certain plane and the beam waist, the wavelength λ, and the M^2 factor, we can calculate the beam waist radius w_0 by modifying Eq. (4.1) to

$$w_0 = \frac{w(z)}{2^{0.5}} \left\{ 1 + \left[1 - \left(\frac{2\lambda z M^2}{\pi w(z)^2} \right)^2 \right]^{0.5} \right\}^{0.5} \qquad (z < z_R) \tag{4.10}$$

or

$$w_0 = \frac{w(z)}{2^{0.5}} \left\{ 1 - \left[1 - \left(\frac{2\lambda z M^2}{\pi w(z)^2} \right)^2 \right]^{0.5} \right\}^{0.5} \qquad (z > z_R) \tag{4.11}$$

The situation for $z > z_R$ and $z < z_R$ is different, as is explained in Fig. 4.2. Beam 1 and beam 2 have their waists w_{01} and w_{02} located at the same position and their size $w(z)$ at z are the same. But their waist sizes w_{01} and w_{02} are very different, because $z < z_R$ for beam 1 and $z > z_R$ for beam 2. When $z = z_R = \pi w_0^2/(\lambda M^2)$, $w(z) = 2^{0.5} w_0$, term $2\lambda z M^2/[\pi w(z)^2] = 1$, Eqs. (4.10) and (4.11) become identical. If we do not know whether $z > z_R$ or $z < z_R$ for a beam, we have to use both Eqs. (4.10) and (4.11) to calculate two values for w_0, then insert these two values of w_0 into Eq. (4.1) to see which value of w_0 will result in a $w(z)$ that matches the measured result.

In most cases encountered, we know neither the beam waist radius w_0, nor the beam waist location, nor the M^2 factor; we only know the wavelength λ. Then, the practically best way is simply to move the beam profiler along the beam propagation axis to find the beam waist radius w_0, calculate the Rayleigh range z_R for this

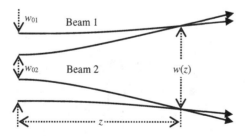

Fig. 4.2 Two Gaussian beams with the same waist locations and same sizes at z can have very different waist sizes

beam waist, then measure the beam radius $w(z)$ at far field with distance $z \gg z_R$. The M^2 factor can be calculated from Eq. (4.4) using the measured w_0, $w(z)$ and z.

Here we would like to note that the ISO 11146 [1] procedure specifies the method of measuring the M^2 factor. The ISO procedure requires that 10 beam sizes along the beam propagation axis are measured. Among the 10 measurements, 5 are around the beam waist and 5 are at least two Rayleigh ranges distance away. For a collimated laser diode beam with 1.5 mm waist radius, 0.67 μm wavelength, and $M^2 = 1.1$, the Rayleigh range can be found from Eq. (4.3) to be $z_R \approx 9.6$ m, twice the Rayleigh range makes 19.2 m, which is a long distance inside a lab. On the other hand, for an unmanipulated laser diode beam, the Rayleigh range is only several microns, a commonly used beam profiler cannot be positioned within a few microns to the laser diode facet, and the distance between the laser diode facet and the beam profiler sensing surface cannot be determined to submicron accuracy, only coarse measurement can be performed.

If we want to characterize the unmanipulated beam of a laser diode, we can use a high quality lens(es) to collimate the beam and measure the waist radius of the collimated beam, then back calculate the waist radius and divergence of the unmanipulated beam, assuming the collimating lens focal length and laser wavelength are known. Such a characterization is easier to perform and the results are more accurate than directly characterizing the unmanipulated beam.

4.1.3 Beam Far Field Divergence Measurement

For a collimated or focused laser diode beam, once we have found the beam waist location z and measured the beam radius $w(z)$ at far field, the beam far field divergence θ can be calculated using Eq. (4.4). There is another way to measure the far field divergence of a laser diode beam without the need to find the beam waist location. Figure 4.3 shows the schematic of a setup for measuring the far field divergence of a collimated laser diode beam. In the setup, a high quality lens(es) with known focal length f is used to focus the beam. For a typical collimated laser diode beam, the Rayleigh range z_R is several meters, a typical lens focal length is

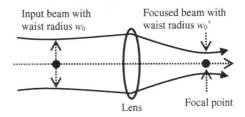

Input beam with waist radius w_0

Focused beam with waist radius w_0'

Lens

Focal point

Fig. 4.3 Schematic of a setup for measuring the far field divergence of a laser diode beam

~ 10 mm, we have $z_R/f \gg o/f - 1$. Combining Eqs. (4.4) and (4.6) to eliminate w_0, we obtain

$$\theta = \frac{w_0}{z_R} = \frac{w_0{}'}{f} \qquad (4.12)$$

Equation (4.12) tells us that we can find the far field divergence θ of a collimated laser diode beam by focusing the beam and measuring the waist radius of the focused beam.

4.1.4 Astigmatism Measurement

Most laser diodes have their beam waists in the slow axis direction located several microns behind the beam waists in the fast axis direction, as shown in Fig. 2.1; although there was a opposite case reported, the inherent astigmatism of laser diodes. Several microns is a small value to measure. The elliptical shape and the large divergence of laser diode beams make the measurement even more complex. If the measurement is not performed carefully, the measurement results can contain large errors or even be erroneous. In this section, we analyze one measurement example.

4.1.4.1 Measurement Setup

The schematic of a measurement setup is shown in Fig. 4.4. The beam of a laser diode to be measured is first collimated by lens 1 and then focused by lens 2. A beam profiler located around the second focal plane of lens 2 monitors the focused beam. The optical magnification ratio of the combination of lens 1 and lens 2 should be large, so that the astigmatism of the focused beam is large and the

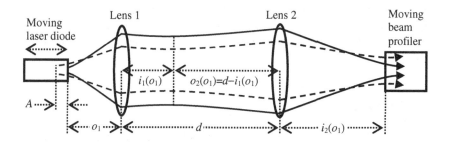

Fig. 4.4 Schematic of a setup for measuring the astigmatism. The *solid* and *dash curves* are for the beam in the fast and slow axis, respectively. The drawing is not to the *right* proportion for illustration purpose

measurement will be easier. Here, we choose the focal lengths of the two lenses to be $f_1 = 5$ mm and $f_2 = 50$ mm, respectively. Then, the setup has a longitudinal magnification of $m_L = (50 \text{ mm}/5 \text{ mm})^2 = 100$. Lens 1 must have a large numerical aperture so that it does not truncate the beam. Otherwise, the beam truncation may cause a focal shift and a large error to the astigmatism measurement result.

In Fig. 4.4, o_1 is the distance between the waist of the laser diode beam and the principle plane of lens 1, $i_1(o_1)$ is the distance between the principle plane of lens 1 and the waist of the beam output from lens 1, $i_1(o_1)$ is a function of o_1 defined by Eq. (4.5). d is the spacing between the principle planes of lens 1 and lens 2. $o_2(o_1)$ is the distance between the principle plane of lens 2 and the waist of the beam input to lens 2, since

$$o_2(o_1) = d - i_1(o_1) \tag{4.13}$$

$o_2(o_1)$ is a function of o_1 in this setup. $i_2(o_1)$ is the distance between the principle plane of lens 2 and the waist of the beam output from lens 2. Since $i_2(o_1)$ is a function of $o_2(o_1)$ defined by Eq. (4.5) and $o_2(o_1)$ is a function of o_1 defined by Eq. (4.13), $i_2(o_1)$ is also a function of o_1.

There are two measurement methods:

1. Moving diode method

The beam profiler is positioned at the focal plane of lens 2. The laser diode is moved back and forth along the optical axis of the setup around the focal point of lens 1. When the beam profiler sees a focused spot in the fast axis direction, record the laser diode position. When the beam profiler sees a focused spot in the slow axis direction, record the laser diode position again. The distance between these two positions of the laser diode is the astigmatism.

2. Moving profiler method

Either the beam waist of the laser diode in the fast axis direction or in the slow axis direction is positioned at the focal point of lens 1. The beam profiler is moved back and forth along the optical axis of the setup around the focal plane of lens 2. When the beam profiler sees a focused spot in the fast axis direction, record the beam profiler position. When the beam profiler sees a focused spot in the slow axis direction, record the beam profiler position again. The distance between these two positions of the beam profiler divided by the magnification of the setup is the astigmatism.

4.1.4.2 Error Analysis

There are pitfalls in the measurement. These pitfalls can cause large measurement errors and must be analyzed. Consider a laser diode with 0.67 μm wavelength, FWHM divergence of 24° and 9° in the fast and slow axis directions, respectively, $M^2 = 1$ and the beam astigmatism $A = 10$ μm. These two divergent angles can be converted to $1/e^2$ intensity full divergence of 40.8 and 15.3°, respectively, by multiplying a coefficient of 1.7. The beam waist $1/e^2$ radii in the fast and slow axis directions can be found using Eq. (4.4) to be $w_{0F} = 0.6$ μm and $w_{0S} = 1.6$ μm, respectively. The Rayleigh ranges of the laser diode beam in the fast and slow axis directions can be calculated using Eq. (4.3) to be $z_{ROF} = 1.7$ μm and $z_{ROS} = 12$ μm, respectively.

The waist radii in the fast and slow axis directions of the beam output from lens 1 can be found using Eq. (4.6) to be, respectively,

$$w_{1F}(o_1) = \frac{w_{0F}}{\left[\left(\frac{o_1}{f} - 1\right)^2 + \left(\frac{z_{ROF}}{f}\right)^2\right]^{0.5}} \tag{4.14}$$

$$w_{1S}(o_1) = \frac{w_{0S}}{\left[\left(\frac{o_1}{f} - 1\right)^2 + \left(\frac{z_{ROS}}{f}\right)^2\right]^{0.5}} \tag{4.15}$$

$w_{1F}(o_1)$ and $w_{1S}(o_1)$ as functions of o_1.

The Rayleigh ranges in the fast and slow axis directions of the beam output from lens 1 can be found using Eq. (4.3) to be, respectively,

$$z_{R1F}(o_1) = \frac{\pi w_{1F}(o_1)^2}{\lambda} \tag{4.16}$$

$$z_{R1S}(o_1) = \frac{\pi w_{1S}(o_1)^2}{\lambda} \tag{4.17}$$

$z_{R1F}(o_1)$ and $z_{R1S}(o_1)$ are functions of o_1 since $w_{1F}(o_1)$ and $w_{1S}(o_1)$ are functions of o_1. The image distance in the fast and slow axis direction of the beam output from Lens 2 can be found using Eq. (4.5) to be, respectively,

$$i_{2F}(o_1) = \frac{o_{2F}(o_1)\left[\frac{o_{2F}(o_1)}{f_2} - 1\right] + \left[\frac{z_{R1F}(o_1)}{f_2}\right]^2}{\left[\frac{o_{2F}(o_1)}{f_2} - 1\right]^2 + \left(\frac{z_{R1F}(o_1)}{f_2}\right)^2} \tag{4.18}$$

$$i_{2S}(o_1) = \frac{o_{2S}(o_1)\left[\frac{o_{2S}(o_1)}{f_2} - 1\right] + \left[\frac{z_{R1S}(o_1)}{f_2}\right]^2}{\left[\frac{o_{2S}(o_1)}{f_2} - 1\right]^2 + \left(\frac{z_{R1S}(o_1)}{f_2}\right)^2} \tag{4.19}$$

Combining Eqs. (4.13)–(4.19), we can plot $i_{2F}(o_1)$ and $i_{2S}(o_1 + A)$ in Fig. 4.5 for three d values, respectively.

From Fig. 4.5a, we can see that for $d = f_1 + f_2 = 55$ mm, both $i_{2F}(o_1)$ and $i_{2S}(o_1 + A)$ are straight lines. Δo_1 or Δi_2 can be found at any locations for the laser diode or the beam profiler, respectively. After performing some numerical

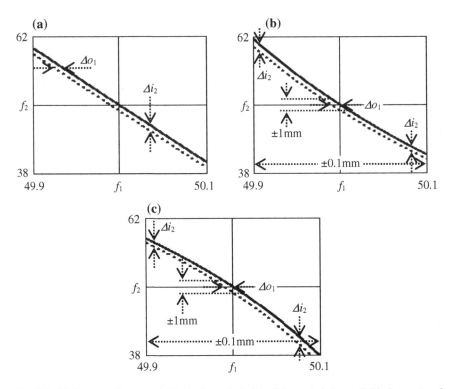

Fig. 4.5 *Horizontal axis o_1 (mm). Vertical axis $i_{2F}(o_1)$ (solid curve), $i_{2S}(o_1 + A)$ (dash curve)* and the unit is mm. **a** $d = f_1 + f_2 = 55$ mm. **b** $d = 20$ mm. **c** $d = 100$ mm

evaluation on Eqs. (4.18) and (4.19), we can find that $\Delta o_1 = 10$ μm $= A$ and $\Delta i_2 = 1$ mm $= m_L \times A = 100 \times 10$ μm. Therefore, the measurement is relatively easier to perform and the error can be kept small.

From Fig. 4.5b, we can see that for $d = 20$ mm $\neq f_1 + f_2 = 55$ mm, both $i_{2F}(o_1)$ and $i_{2S}(o_1 + A)$ are no longer straight lines, but are curves. When using the moving diode method, it is not too difficult to position the beam profiler within ±1 mm to the focal plane of lens 2. The nonlinearity of the two curves is not obvious within this range, as shown in Fig. 4.5b. After performing some numerical evaluations on Eqs. (4.18) and (4.19), we can find that $\Delta o_1 \approx 10$ μm with a small error of ~0.2 μm. When using the moving profiler method, it is not easy to position the laser diode emission facet within ±0.1 mm to the focal position of lens 1, since the laser diode chip is tiny and usually hides inside the cap. But, assume that we can still manage to position the laser diode emission facet within ±0.1 mm to the focal position of lens 1. The nonlinearity of the two curves within this range is obvious, as shown in Fig. 4.5b. After performing some numerical evaluations on Eqs. (4.18) and (4.19), we can find that $\Delta i_2 \approx 1.36$ mm and ≈ 0.77 mm at $o_1 = 4.9$ mm and 5.1 mm, respectively. The measured astigmatism is 1.36 mm/$m_L = 13.6$ μm and 0.77 mm/$m_L = 7.7$ μm, which is a large error of 36 and 23 %, respectively.

The actual position error of the laser diode can easily exceed ±0.1 mm, the measurement error can be even larger.

Similarly, from Fig. 4.5c, we can see that for $d = 100$ mm $\neq f_1 + f_2 = 55$ mm, both $i_{2F}(o_1)$ and $i_{2S}(o_1 + A)$ are curves. The measurement error of moving diode method will be ~ 0.26 μm assuming a generous positioning error of ±1 mm for the beam profiler. The measurement error of moving profiler method will be up to 47 % assuming a very tight positioning error of ±0.1 mm to the laser diode.

In conclusion, when we measure laser diode beam astigmatism using the two lens setup shown in Fig. 4.4, we need set the lens spacing to $d = f_1 + f_2$ and using moving diode method whenever possible.

4.2 Spectral Property Characterization: Wavelength Measurement

The spectral properties of laser diodes are described by the wavelength and line-width, same as for the other types of lasers. However, for most laser diodes, the wavelength is not stable and the linewidth is large. The techniques for measuring the spectral properties of laser diodes do not need to have high accuracy and are sometimes different from the techniques for measuring the spectral properties of other types of lasers.

Most lasers have fixed wavelength, there is no need to measure their wavelength. For example, a red He–Ne laser has a fixed wavelength of 632.8 nm. Wavemeters were first developed to measure the wavelength of tunable lasers, such as dye lasers. Commercial laser wavemeters in market now can be categorized into two types: Fizeau wavemeter and Michelson interferometer wavemeter. These commercial wavemeters are accurate, reliable, and can be used to measure the wavelength of DFB laser diodes, DBR laser diodes, and external cavity tunable laser diodes. The manuals of these wavemeters should provide enough information about how to operate these instruments. Here we describe the working principles of these two types of wavemeters.

4.2.1 Fizeau Wavemeter

This type of wavemeters uses a Fizeau wedge, either air wedge or glass wedge, to generate interference fringes from the laser beam to be measured and use a CCD linear array to detect the fringes, as shown in Fig. 4.6. If a glass wedge is used, the measurement result must be calibrated using the glass dispersion data already saved in the data processing software. The changing of wavefront curvature of the beam being measured will affect the measurement accuracy. Therefore, the beam being measured must first be coupled into a single mode fiber and the beam output by the

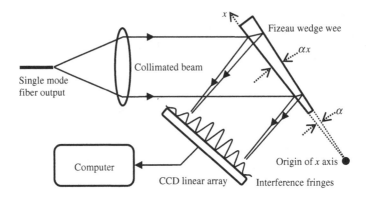

Fig. 4.6 Schematic of a Fizeau pulsed laser wavemeter, the wedge thickness is αx

fiber must be collimated. Fizeau wavemeter has no moving parts; the measurement can be completed as long as the CCD can catch the signal. Therefore, Fizeau wavemeters can measure the wavelength of both continuous and pulsed lasers. The shortcoming of Fizeau wavemeters is that the wavelength range a linear sensing array can cover is limited which limits the measurement range of Fizeau wavemeters.

Fizeau wavemeter was invented by Snyder [2] and is the most widely used wavemeter nowadays. An Internet search with keywords "Fizeau wavemeter" can find several publications about the design and testing of a Fizeau wavemeter, for example Refs. [3, 4]. The data processing algorithm [5] developed by Snyder et al. for a Fizeau wavemeter is simple and effective, and will be explained in Sect. 4.2.4.

The working principle of a Fizeau wavemeter is explained below. The wavelength λ to be measured and the period p of the interference fringe are related by

$$\lambda = F(\alpha)p \tag{4.20}$$

where $F(\alpha)$ is a function of wedge angle α and is known by calibration. Using a CCD linear array with 1024 or more pixels and the data processing algorithm developed for Fizeau wavemeters [5], p can be measured with an error of $<10^{-4}$, we call this wavelength λ_1, the true wavelength λ must fall inside the range of

$$\lambda_1 - 10^{-4}\lambda_1 < \lambda < \lambda_1 + 10^{-4}\lambda_1 \tag{4.21}$$

A measurement error of $<10^{-4}$ is adequate for measuring the wavelength of most laser diodes, but is not enough for measuring the wavelength of most other types of lasers; therefore, commercial Fizeau wavemeter must move one step further. For any point x on the Fizeau wedge, the optical path difference $D(x)$ between the two rays that interfere each other is known by calibration. We have relation

$$\lambda = \frac{D(x)}{m + \Delta m} \tag{4.22}$$

where m is an unknown interference order and Δm is a fraction of interference order that can be measured with an error $<10^{-4}$ same as measuring p. $D(x)$ is proportional to the local thickness αx of the wedge. We can plug many trial values of m into Eq. (4.22) and find many values for λ. It can be shown that when the local thickness is properly chosen, there is only one value of m that will lead to a λ falling inside the range of Eq. (4.21), this λ is the measurement result. Since $m \sim 1000$, $m + \Delta m$ contain an error of $<10^{-7}$, the λ found from Eq. (4.22) will also contain an error of $<10^{-7}$, which is the final measurement error of most commercial Fizeau wavemeters.

Below is a conceived measurement to illustrate how the measurement process is going. Assume a laser has a true wavelength of 1000.000 nm, but we do not know and want to measure it. From Eq. (4.21), we find that the wavelength is somewhere inside the range of 1000 nm $\pm 10^{-4}$ nm. Assuming at certain x on the wedge, $D(x) = 2$ mm and $\Delta m = 0 \pm 10^{-4}$. If we guess $m = 2000$, we can find from Eq. (4.22) that $\lambda = 2$ mm/$(2000 \pm 10^{-4}) \approx (2$ mm/2000$)(1 \pm 10^{-4}/2000) = 1000$ nm $\pm 5 \times 10^{-8}$ nm, this result falls inside the range of 1000 nm $\pm 10^{-4}$ nm, is the correct result and has high accuracy. If we guess $m = 1999$, from Eq. (4.22) we find $\lambda = 2$ mm/$(1999 \pm 10^{-4}) \approx (2$ mm/1999$)$ $(1 \pm 10^{-4}/1999) = 1000.5$ nm $\pm 5 \times 10^{-8}$ nm, this result falls outside the range of 1000 nm $\pm 10^{-4}$ nm and is erroneous. Similarly, if we guess $m = 2001$, we will find an erroneous result of $\lambda = 999.5$ nm $\pm 5 \times 10^{-8}$ nm.

As shown in Fig. 4.6, the wavemeter is so aligned that the two rays meeting on the CCD array to interfere are from the same incident ray on the Fizeau wedge. If there are any aberrations on the wavefront of the beam to be measured, the same aberrations will be carried by the two rays and will cancel each other when the two rays interfere. Such an alignment can reduce the sensitivity on the wavefront quality and reduce the measurement error. The commercial Fizeau wavemeters often have a temperature control device inside the case to reduce the measurement error caused by thermal expansion, and a calibration He–Ne laser inside the case for occasional self calibration.

4.2.2 Michelson Interferometer Wavemeter

This type of wavemeter uses a moving arm Michelson interferometer to generate interference fringes from the laser beam to be measured, as shown Fig. 4.7 [6]. The laser beam to be measured is split into two by a beamsplitter. The two beams propagate in two optical arms and are then recombined by the beamsplitter to interfere. As the length of one optical arm changes, the intensity of the interference varies periodically. A single element sensor monitors the intensity of the

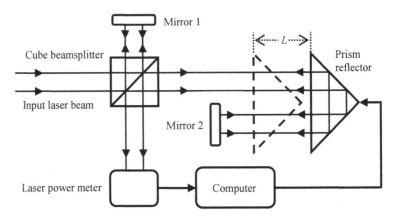

Fig. 4.7 Schematic of a Michelson interferometer measurement setup

interference signal and counts the periods. The wavelength λ being measured can be calculated from the number $m + \Delta m$ the sensor counting and the optical arm moving distance L by

$$\lambda = \frac{4L}{m + \Delta m} \qquad (4.23)$$

where m is an integer and Δm is a fraction order. For an arm moving distance of $L \sim 200$ mm, m $\sim 10^6$ and Δm can be determined to an accuracy of <0.1. The wavelength calculated from Eq. (4.23) has an error of $\sim 10^{-7}$.

Michelson interferometer wavemeters have a moving arm; a measurement takes at least several seconds to complete and cannot measure the wavelength of pulsed lasers, which is the disadvantage. The advantage is that single element sensors available in market can cover wider wavelength range than linear arrays. The commercial Michelson interferometer wavemeters often have a temperature control device inside the case to reduce the measurement error caused by thermal expansion and a calibration He–Ne laser inside the case for occasional self calibration.

4.2.3 Lab Wavelength Measurement Setup

In some applications, the wavelength of laser diodes needs be measured. The wavelength of the same type of laser diodes can have a few nanometers variation caused by manufacturing tolerance and the wavelength of the very same laser diode shifts at a rate of about 0.2–0.3 nm/°C. These commercial wavemeters described in the above sections may have trouble measuring the wavelengths of laser diodes, because the coherent length of laser diode beams may be shorter than the optical path difference that is used to generate interference fringes in these wavemeters.

On the other hand, these commercial wavemeters are expensive. The unit price is from ∼$10 to ∼$20 K, their measurement accuracy of higher than 0.001 nm can be excessive for measuring the wavelengths of laser diodes with a linewidth of ∼0.1 nm, except for DFB, DBR, and external cavity laser diodes.

In this section, we describe two types of simple, low cost wavelength measurement setup with measurement accuracy of ∼0.1 nm, adequate for measuring laser diode wavelength. We can utilize some commonly available lab equipment and materials to build these simple setups.

4.2.3.1 Young's Double Slit Wavelength Measurement Setup

The schematic of such a setup is shown in Fig. 4.8 [7]. The double slits are made by carving the metal coating on a mirror. The width of the two slits is $a \approx 0.01$ mm, the spacing between the two slits is about $d \approx 0.5$ mm, the length of the slits can be a couple of mm. The two slits will be calibrated before taking measurement, so they do not need be made accurately. The back surface of the mirror must be an optical surface. A beam slitter or an optical glass window substrate or even a microscope slide combines an expanded He–Ne laser beam for calibration and the collimated laser diode beam to be measured. The two beams are incident on the double slits and generate interference fringes on a CCD linear array, and a cylindrical lens can be placed in front of the CCD array to focus the light in the direction of the slits onto the CCD array in case the intensity of light is too low. The distance between the slits and the CCD array is about $D \approx 200$ mm.

The diffraction intensity angular distribution of one slit is given by

$$I(\theta) \sim \sin c \left[\frac{a\pi}{\lambda} \sin(\theta) \right]^2 \qquad (4.24)$$

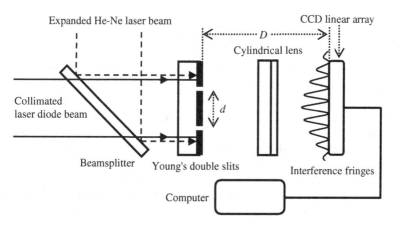

Fig. 4.8 Schematic of a simple setup using a Young's double slits to measure laser diode wavelength

The FWHM angular diameter of $I(\theta)$ is $\theta_{FWHM} \approx 0.06$ Radian assuming $\lambda \approx 0.7$ μm. The FWHM linear diameter on the CCD array is $D \times \theta_{FWHM} \approx 12$ mm enough to cover a 10.24 mm-long CCD array assuming the CCD has 1024 pixels of 10 μm size. If we use a different size CCD linear array, we can adjust D so that the diffraction pattern comfortably covers the array. The diffraction profiles of the two slits are only ~ 0.5 mm apart. The add-up of the two intensity profiles on the CCD array can still be approximated by Eq. (4.24). The numbers of the interference fringes on the CCD array is $N = d \times \theta_{FWHM}/\lambda \approx 43$. Any values between ~ 10 and ~ 100 are fine for N.

The sampling data from the CCD is sent to a computer for processing. When inventing the Fizeau wavemeter, Snyder also developed an algorithm that can calculate the period of interference fringes to a high accuracy [5]. This algorithm can be effectively used here as well. Analysis and tests results show that fringe period can be calculated to an accuracy of 10^{-4} with a 1024 pixel CCD array [8]. When performing measurement using this double slits setup, we first block the laser diode beam, only measure and calculate the fringe period P_H of the He–Ne laser; this procedure is to calibrate the setup. Then, we block the He–Ne laser beam, only measure and calculate the fringe period P_L of the laser diode beam. The unknown wavelength of the laser diode can be calculated with an error of $\sim 10^{-4}$ to be

$$\lambda_L = 632.8 \, \text{nm} \frac{P_L}{P_H} \qquad (4.25)$$

The thermal expansions of the glass substrate on which the Young's double slits are made and the CCD array are about $\sim 10^{-6}/°C$. The temperature variation of the CCD array can be tens of °C since it consumes electrical power. Before measuring, the CCD array must be operated for at least several minutes so that its temperature change is slowed down. The time interval of measuring the He–Ne laser and the laser diode should be kept as small as possible.

4.2.3.2 Glass Wedge Wavelength Measurement Setup

The schematic of the measurement setup is shown in Fig. 4.9. Both the He–Ne laser beam for calibration and the laser diode beam to be measured need be collimated and expanded to over 10 mm size, which is large enough to cover the ~ 10 mm length of the CCD linear array used. Another purpose of expanding the beams to a large size is that the interference fringe period, upon which the wavelength is calculated, is proportional to the beam divergence in this setup. Larger beams have smaller divergences and smaller divergence variations as the beams propagate, and the measurement results will be more accurate. A beamsplitter or an optical window substrate or even a microscope slide combines the two expanded and collimated beams. The two beams are incident on a selected optical glass window substrate. The reflected beams from the two surfaces of the glass substrate generate

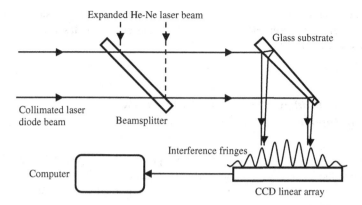

Fig. 4.9 Schematic of a simple setup using a glass wedge to measure laser diode wavelength and linewidth

interference fringes on a CCD linear array. A computer takes the CCD output data to calculate the wavelength.

This setup is very similar to a Fizeau wavemeter. However, there are two differences:

1. A Fizeau wedge is not an off-shelve optical element, it must be specially fabricated and costs hundreds of dollars. In this setup, we use a low-cost glass substrate instead of a Fizeau wedge.
2. There are no special requirements on the incident angle of the beams on the glass substrate and the position of the linear CCD array relative to the glass substrate, all because the target measurement accuracy of this setup is low, only $\sim 10^{-4}$. The alignment of the setup is easy.

Every glass window substrate has a random wedge angle between the two surfaces. Those glass window substrates with 1λ surface quality usually have a few arc minutes wedge angle. Shine an expanded He–Ne laser beam on several optical glass window substrates, respectively, and use a piece of white paper to intercept the reflected beam and observe it; you will have a good chance to find that one or two substrates can generate nice interference fringes. By changing the incident angle of the laser beam on the substrate, the size of fringe period can be changed. If the sensing length of the CCD array can cover over 10 fringes or so, this glass substrate and this incident angle are right for our measurement setup. Actually, we can even find a good glass substrate from several microscope slides. These slides cost less than one dollar each.

Then, sample the He–Ne laser beam and the laser diode beam separately, and use Snyder's algorithm [5] to calculate the fringe periods P_H and P_L of the two lasers, the wavelength of the laser diode can be found with an error of $\sim 10^{-4}$ using Eq. (4.25). Same as in the Young's double slits measurement, the CCD array needs be operated for several minute so that its temperature changing is slow down, and the time interval between measuring the He–Ne laser and the laser diode should be kept as small as possible.

4.2.4 Algorithm for Calculating the Periods of Interference Fringe

Snyder et al. developed an algorithm that can process the fringe data sampled by the CCD array and calculate the fringe period to an accuracy of ~0.1 pixel [5, 8]. Here we provide some additional explanations to help readers understand how the algorithm works.

A digital filter is built and scanning through the fringe sampling data, as shown in Fig. 4.10. The values of all the sampling points inside the range of the left hand half of the filter are added up and multiplied by −1, we call this value F_L. The values of all the sampling points inside the range of the right-hand half of the filter are added up and multiplied by 1, we call this value F_R. The width of the whole filter is first set equally to the estimated width of one fringe, as shown in Fig. 4.10. When the filter is at the (a) position shown in Fig. 4.10, the center line of the filter is at the left-hand side of the peak of one fringe, we have $|F_L| < F_R$ or $F_L + F_R > 0$. When the filter is at the (b) position shown in Fig. 4.10, the center line of the filter is at the right-hand side of the peak of one fringe, we have $|F_L| > F_R$ or $F_L + F_R < 0$. As the filter is scanned through the sampling points, whenever $F_L + F_R$ changes sign from positive to negative, we know the center line of the filter just passes through the peak of one fringe. Similarly, as the filter is scanning through the sampling points, whenever $F_L + F_R$ changes sign from negative to positive, we know the center line of the filter just passes through the valley of one fringe. The actual peak and valley positions of one fringe are often in between two sampling points. This fraction can be calculated by $|F_L|/(|F_L| + F_R)$ with an accuracy of 0.1. For example, if the peak position of one fringe is at the middle of two sample points, the situation is symmetric, we have $|F_L| = F_R$ and $|F_L|/(|F_L| + F_R) = 0.5$.

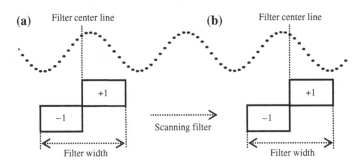

Fig. 4.10 A digital filter scans through the fringes

The approximate peak and valley positions of every fringe can be found by scanning the filter through all the sampling data. Then using the least squares method to fit all these peak and valley positions to reduce any random errors, the fringe period P can be determined to an accuracy of 10^{-4} if 1,000 sampling points are used. It is shown that the best width of the filter is about $0.7P$ [5]. If the filter width first chosen by the estimation is far off from $0.7P$, the filter width needs to be reset to $0.7P$ based on the first time calculation data; then the filter needs to be scanned through the sampling data again to improve the accuracy.

4.3 Spectral Property Characterization: Linewidth Measurement

Most lasers have linewidth much narrower than the linewidth of laser diodes. For example, a commonly used He–Ne laser has a linewidth of ~ 1 GHz or ~ 0.001 nm. Heterodyne interferometer type techniques have the highest measurement resolution and is the right one to measure linewidth of <0.1 GHz. Scanning Fabry–Perot interferometers (SFPIs) have resolutions of about 0.1–1 GHz or about 0.0001–0.001 nm and is widely used to measure laser linewidth or analyze the spectrum of laser light. But the measurement range (free spectral range) of SFPIs is often too small to cover the large linewidth of laser diodes. Diffraction grating-based monochromators usually have resolutions of 10–100 GHz or about 0.01–0.1 nm. This resolution is too low for measuring the linewidth of most other types of lasers, and a little too low for measuring the linewidth of many laser diodes. There are several specially designed and developed commercial spectrum analyzers in market. These analyzers have both high accuracies and wide measurement ranges, and a price tag usually over $20 K.

In this section, we first describe the working principles if gratings and SFPIs, since these devices are most widely used. Then, we describe a lab setup for linewidth measurement.

4.3.1 Diffraction Gratings

Diffraction gratings are the most commonly used diffraction devices and the key elements inside a monochromator. Several different types of diffraction gratings are available in market. The most commonly used gratings are planar reflective gratings. They are a collection of many small, identical, and slit-shaped grooves ruled on a planar high reflective surface. Figure 4.11 shows the schematic of two grooves of such a grating.

The working principle of a grating is explained below. Consider two rays in an incident beam, marked by "Ray 1" and "Ray 2," incident on the identical points of

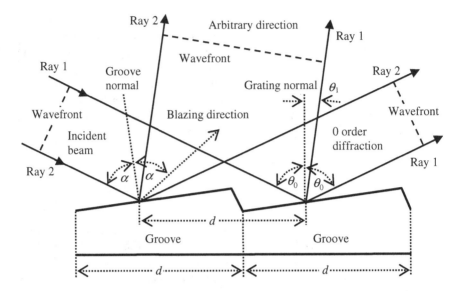

Fig. 4.11 Two grooves of a grating diffract a beam

the two grooves, respectively, as shown in Fig. 4.11. Ray 1 and Ray 2 in both the incident beam and the zero-order diffraction beam have the same angle θ_0 about the grating normal, the optical path of Ray 1 and Ray 2 are the same. That means these two rays have constructive interference for any wavelengths in this zero-order diffraction direction. It is noted that the zero-order diffraction direction does not meet the reflection condition since the surface of the grooves are tilted. The intensity of zero-order diffraction can be weak or even zero if the zero-order diffraction direction is much off the reflection direction.

Figure 4.11 also shows an arbitrary direction. The optical path difference between Ray 1 and Ray 2 in this direction is given by $d[\sin(\theta_0) + \sin(\theta_1)]$, where d is the groove period and θ_1 is the angle between this arbitrary direction and the grating normal. The well-known grating equation is

$$d[\sin(\theta_0) + \sin(\theta_1)] = m\lambda \qquad \text{(Constructive interference)}$$
$$d[\sin(\theta_0) + \sin(\theta_1)] = (m + 1/2)\lambda \quad \text{(Destructive interference)} \qquad (4.26)$$

where m is an integer called "diffraction order." The signs of θ_0 and θ_1 are defined as positive at the left-hand side of the grating normal and negative at the right-hand side of the grating normal. For a given λ, the maximum m allowed is $m_{\max} = 2d/\lambda$ obtained by letting $\sin(\theta_0) + \sin(\theta_1) = 2$ in Eq. (4.26). For a given m and θ_0, θ_1 is a function of λ that means different wavelength components in a beam will have constructive interference diffractions in different directions.

The "Blazing direction" marked in Fig. 4.11 is the reflection direction about the groove normal. The "Blazing wavelength" is the wavelength that has constructive interference in the blazing direction. The wavelength with constructive interference in the blazing direction has the maximum intensity since this direction meets the reflection condition. As the wavelength shifting away from the blazing wavelength, the direction for constructive interference moves sway from the blazing direction, the intensity of the diffracted beam drops.

Any other rays in the incident beam can be grouped in pairs similar to the pair of Ray 1 and Ray 2 and analyzed in the same way. Thereby the whole incident beam and diffracted beam have the same characteristics of Ray 1 and Ray 2.

The angular dispersion resolution or the resolving power of a grating is defined by $d\theta_1/d\lambda$ and can be obtained by differentiating Eq. (4.26) with θ_0 being a constant. The result is

$$\frac{d\theta_1}{d\lambda} = \frac{m}{d\cos(\theta_1)} \tag{4.27}$$

Equation (4.26) shows that when $m = 0$, $\theta_1 = -\theta_0$. Equation (4.27) shows that when $m = 0$, $d\theta_1/d\lambda = 0$, which means the zero-order diffraction has no dispersion effect. Large $d\theta_1/d\lambda$ is often desired and can be obtained by the use of a small d, a large m and a large θ_1 close to the practical limit of 90°. Commonly used diffraction gratings have a groove density from 300/mm ($d \approx 3.33$ μm) to 2400/mm ($d \approx 420$ nm). The corresponding dispersion resolution for $m = 1$ is from $d\theta_1/d\lambda = 4 \times 10^{-4}$ rad/nm to $d\theta_1/d\lambda = 3.4 \times 10^{-3}$ rad/nm.

Grating efficiency is the percentage of incident monochromatic light that is diffracted into the desired order, and is determined by the groove shape, angle of incidence, and the reflectance of the coating. Grating efficiency is probably more complex to calculate than expected and should be provided by the grating vendors.

Groove density, blazing wavelength, resolving power, wavelength range, and grating efficiency are the main parameters describing a diffraction grating. Richardson Gratings (http://www.gratinglab.com) is the largest grating manufacturer in the world, their "Diffraction Grating Handbook" is a great technical resource for gratings, readers can go to this website, http://www.gratinglab.com/Information/Handbook/Handbook.aspx, asking for a free copy.

4.3.2 Monochromators

Monochromators are the most commonly used instruments for optical spectrum analysis. Various types of monochromators have been developed. Figure 4.12 shows the schematic of a simple monochromator. The key device in a monochromator is the diffraction grating. A diffraction grating can spatially disperse a polychromatic laser beam into its monochromatic components. The laser beam under measurement is incident through an entrance slit, collimated by two concave mirrors, diffracted by a grating, and focused by another concave mirror on an exit slit. A laser power meter measures the power of the beam passing through the exit

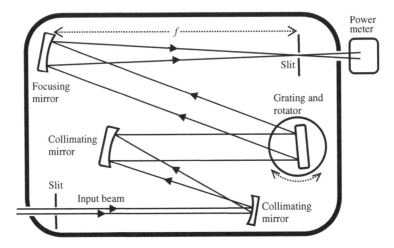

Fig. 4.12 Schematic of a monochromator

slit. The grating is mounted on a rotator. The positions of the two slits are fixed. For any given grating orientation and the laser beam incident angle, the diffraction angle θ_0 and θ_1 are known. The wavelength λ of the diffracted beam passing through the exit slit can be calculated using Eq. (4.26). By rotating the grating and recording the power measured by the laser power meter as a function of the corresponding λ, we can measure the spectrum. The widths of the two slits; Δ_{s1} and Δ_{s2}; are adjustable. The three mirrors image the entrance slit on the exit slit. Δ_{s1}', the width of the image of Δ_{s1}, is proportional to Δ_{s1}. The measurement resolution R of a monochromator is given by Eq. (4.28)

$$R = \frac{\Delta_s}{f} \frac{d\lambda}{d\theta_1} \qquad (4.28)$$

where Δ_s equals to the larger of Δ_{s1}' and Δ_{s2}' f is the focal length of the focusing mirror as shown in Fig. 4.12, and $d\lambda/d\theta_1$ is the inverse of the angular resolution of the grating used. Equation (4.28) shows that reducing the widths of the two slits can increase the measurement resolution. However, the slits must be wide enough to allow enough laser power passing through for measurement. The grating rotation angular resolution also affects the measurement resolution of a monochromator. A monochromator often comes with a few gratings with different groove densities. The measurement resolution and range of the monochromator can be changed by changing the grating.

The resolution of commonly used monochromators is about 0.1 nm that is a little too low for measuring the linewidth of laser diodes. There are some higher resolution monochromators available in market. For example, Horiba (http://www.horiba.com) iHR550 monochromator has a resolution of 0.025 nm, and Spectral

Products (http://www.spectralproducts.com) DK480 1/2 meter Monochromator has
a resolution of 0.03 nm. These monochromators can be used to measure the linewidth of laser diodes.

4.3.3 Scanning Fabry–Perot Interferometers

SFPIs are probably the most widely used instrument for analyzing the spectrum and
measuring the linewidth of lasers. Although their measurement ranges can be too
small and their resolutions can be excessive for measuring the linewidth of most
laser diodes, we still devote one section here to describe the working principle of
SFPIs.

An SFPI consists of two slightly wedged transparent plates with flat surfaces as
shown in Fig. 4.13. The two inner surfaces of the plates are set parallel to each
other, usually polished to a flatness of better than $\lambda/8$, and are high reflection coated.
The two outer surfaces are also optically polished, antireflection coated, and have
an angle between them and the two inner surfaces. So that reflections of the two
outer surfaces cannot interfere with the reflections of the two inner surfaces. The
distance D between the two inner surfaces can be adjusted by a piezoelectric device.
The medium between the two plates is usually air with a unit refractive index. The
laser beam under measurement is collimated by a lens system and incident on the
SFPI, multiple reflections take place at the two inner surfaces, as shown in
Fig. 4.13, and produce a series of transmitted beams whose amplitudes fall off
progressively. The beams transmitted through the SFPI are focused by another lens
system onto a laser power meter.

From Fig. 4.13 we can see that the phase difference between two successive
transmitted beams is

$$\Delta\phi = \frac{4\pi D}{\lambda} \tag{4.29}$$

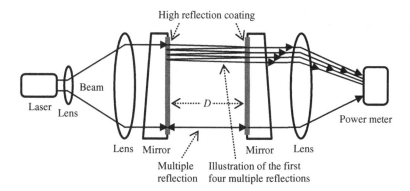

Fig. 4.13 Schematic of a scanning Fabry–Perot interferometer

Interference occurs among the amplitudes of the transmitted beams. The intensity of the combination of all the transmitted beams is given by Wyant [9]

$$
\begin{aligned}
I(\phi) &= I_0 \left| t^2 \sum_{k=0}^{\infty} r^{2k} \exp(-ik\Delta\phi) \right|^2 \\
&= I_0 \left| \frac{t^2}{1 - r^2 \exp(-i\Delta\phi)} \right|^2 \\
&= I_0 \frac{(1 - r^2)^2}{1 + r^4 - 2r^2 \cos(\Delta\phi)} \\
&= I_0 \frac{1}{1 + F \sin\left(\frac{\Delta\phi}{2}\right)^2} \qquad (4.30)
\end{aligned}
$$

where I_0 is the intensity of the beam before incident on the SFPI, $r < 1$ and $t = (1 - r^2)^{0.5}$ are the amplitude reflection and transmission coefficient of the two inner surfaces of the SFPI, respectively, and $F = 4r^2/(1 - r^2)^2$ is a parameter describing the SFPI called "Fineness".

The FWHM of $I(\phi)$ can be found by letting $I(\phi) = 0.5I_0$ in Eq. (4.30) and solving for $\Delta\phi$, the result is

$$
\begin{aligned}
\Delta\phi_{\mathrm{FWHM}} &= 4 \sin^{-1}\left(\frac{1}{F^{0.5}}\right) \\
&= 4 \sin^{-1}\left(\frac{1 - r}{2r}\right) \qquad (4.31) \\
&\approx 2 \frac{(1 - r)}{r}
\end{aligned}
$$

where the approximation is taken because r is usually close to 1. Equation (4.31) shows that the fringe width of $I(\phi)$ reduces as r is increased, because larger r results in more reflections between the two inner surfaces of the SFPI and more beams being involved in interference. The FWHM of $I(\phi)$ in term of wavelength can be found from Eq. (4.29) as

$$
\begin{aligned}
\Delta\lambda_{\mathrm{FWHM}} &= \frac{4\pi D}{\Delta\phi_{\mathrm{FWHM}}} \\
&= \frac{\pi D}{\sin^{-1}\left(\frac{1}{F^{0.5}}\right)} \qquad (4.32)
\end{aligned}
$$

Constructive interference occurs when $\Delta\phi = 2\,m\pi$. From Eq. (4.29) we have

$$\lambda_m = \frac{2D}{m} \qquad (4.33)$$

where λ_m are the wavelengths at which the peak of $I(\phi)$ appears and m is an integer. It's noted that for any given D, there are many λ_m since m can take many values. The spacing between two adjacent peak wavelengths can be obtained by differentiating Eq. (4.33) with respect to m, eliminating m and letting $\Delta m = 1$, the result is

$$\Delta\lambda = \frac{\lambda^2}{2D} \qquad (4.34)$$

$\Delta\lambda$ is known as the "Free spectral range" (FSR) of an SFPI.

Equation (4.30) is plotted in Fig. 4.14 with λ being the variable for several D and F values, and with $I_0 = 1$. We can see that a SFPI acts like a bandpass comb. For a given D value, increasing the values of F can reduce the bandwidth $\Delta\phi_{\text{FWHM}}$ as shown in Fig. 4.14a. Most SFPIs have an $r > 0.95$ to reduce the $\Delta\phi_{\text{FWHM}}$. The measurement resolution of an SFPI is limited by $\Delta\phi_{\text{FWHM}}$. For a given F value, increasing the value of D will reduce the $\Delta\phi_{\text{FWHM}}$ at the cost of reducing the FSR $\Delta\lambda$. When D is scanned, the transmitted wavelength is scanned according to Eq. (4.33), the laser power meter measures the outputs of the SFPI as a function of wavelength and thereby measures the laser spectrum.

To properly measure the linewidth of a laser beam, some preknowledge about the linewidth is required. Figure 4.15 shows the transmission bands of a SFPI and three laser lines with different linewidths. The left-hand side line has a width not much larger than the bandwidth of the SFPI, scanning the SFPI band over this laser line will lead to a low accuracy measurement result. The central line has a width larger than the FSR of the SFPI, will transmit through two bands of the SFPI simultaneously. The laser power meter cannot tell the power it receives is from which band and the measurement result will be erroneous. The right-hand side line

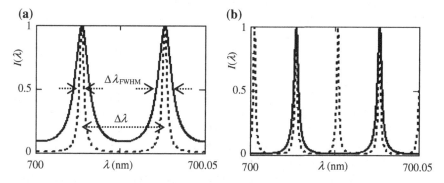

Fig. 4.14 Normalized transmission intensity as a function of wavelength. **a** $D = 10$ mm. *Solid curve $F = 10$, dash curve $F = 100$*. **b** $F = 100$. *Solid curve $D = 10$ mm, dash curve $D = 20$ mm*

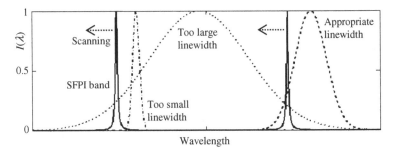

Fig. 4.15 *Solid curves* are the transmit bands of a SFPI. *Dash* or *dot curves* are the three laser lines. The SFPI is only appropriate to measure the linewidth of the *right hand side line*

has a width much larger than the bandwidth of the SFPI, but smaller than the FSR, the measurement result will have a high accuracy.

Commonly used commercial SFPI, for example, the series of SFPIs sold by Thorlabs (http://www.thorlabs.com), has a FSR of about 0.01 nm and a resolution of about FSR/100 = 0.0001 nm. This FSR and resolution are good for measuring the linewidth of typical DFB, DBR, and external cavity laser diodes. But this FSR is too small for measuring the linewidth of most other laser diodes.

4.3.4 Lab Setup for Linewidth Measurement: Glass Flat Setup

When measuring wavelength, an accuracy of higher than 10^{-4} is usually required. When measuring linewidth, an accuracy of higher than 10^{-1} is usually enough. Reference [10] describes using an air-spaced Fizeau wavemeter to measure linewidth by observing the depth of interference fringes. In this measurement, the two beams reflected by the two surfaces of the Fizeau wedge have the same intensities. Such a Fizeau wavemeter is expensive. We can actually use an inexpensive glass flat to perform such a measurement. In our measurement, the two beams reflected by the two surfaces of the glass flat have slightly different intensities. The math involved is slightly more complex, but the measurement principle remains the same.

As explained in Sect. 4.2.3.2, every glass flat has a certain wedge angle between the two surfaces. From several glass flats, we can likely find one flat that has the right wedge angle to generate interference fringes for a CCD linear array to sense. The schematic of a measurement setup is shown in Fig. 4.16a, similar to a Fizeau wavemeter. The fringe number covered by the CCD should be from ~ 10 to ~ 100. The laser beam to be measured is first collimated, then incident on a glass flat. The glass flat splits the incident beam into two beams. There is an optical path difference

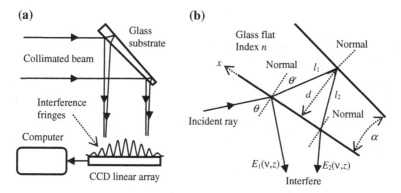

Fig. 4.16 **a** Schematic of a setup for measuring linewidth. **b** The geometry of the optical path difference between the two rays that interfere

$n(l_1 + l_2)$ between the two beam, as shown in Fig. 4.16b. Since the glass wedge angle $\alpha < 1°$, we can have the approximation

$$n(l_1 + l_2) \approx \frac{2nd}{\cos(\theta')}$$
$$= \frac{2n^2d}{\sqrt{n^2 - \sin^2(\theta)}} \qquad (4.35)$$
$$= Bd$$

where $d = \alpha x$ is the glass thickness that is a slow varying function of x, and B is a constant for a given measurement setup. The origin of x axis is at the point where the extension of the two surfaces meets. The amplitude of the two beams reflected by the two surfaces can be described, respectively, by

$$E_1(v, z) = rA(v)e^{2\pi iv\frac{z}{c}}$$
$$E_2(v, z) = r(1 - r^2)A(v)e^{2\pi iv\frac{z}{c}}e^{2\pi iv\frac{Bd}{c}} \qquad (4.36)$$

where r and $(1 - r^2)^{0.5}$ are the amplitude reflectivity and transmittance of the glass surface, respectively, $A(v)$ is the linewidth profile to be measured, v is the frequency of the laser, c is the velocity of light in a vacuum, and z is the spatial variable in the propagation direction of the beam. In Eq. (4.36), we use v as the variable instead of λ to simplify the integration below.

For an uncoated glass flat, the amplitude reflectivity of the glass–air surface is $r = (n - 1)/(n + 1)$. For commonly used optical glasses, n is from 1.4 to 2, and r is from 0.17 to 0.33. The ratio $E_1(v, z)/E_2(v, z)$ is $(1 - r^2) \approx 0.96$, if we use BK7 glass with a refractive index $n = 1.52$. There are actually multireflections between the two surfaces of the glass flat. However, the amplitude of these multireflections drops successively by a factor of $r^2 \approx 0.04$ for BK7 glass, is much weaker and can be neglected.

The intensity of the two beams combined is given by

$$I(v) = |E_1(v,z) + E_2(v,z)|^2 = r^2 |A(v)|^2 \left[1 + p^2 + 2p \cos\left(\frac{2\pi vBd}{c} \right) \right] \quad (4.37)$$

where $p = (1 - r^2)$. The normalized intensity with the linewidth profile being considered is given by

$$J(x) = \frac{\int_{v_0 - \Delta v/2}^{v_0 + \Delta v/2} |A(v)|^2 \left[1 + p^2 + 2p \cos\left(\frac{2\pi vBd}{c} \right) \right] dv}{\int_{v_0 - \Delta v/2}^{v_0 + \Delta v/2} 2|A(v)|^2 dv}$$

$$= \frac{1 + p^2}{2} + \frac{p \int_{v_0 - \Delta v/2}^{v_0 + \Delta v/2} |A(v)|^2 \cos\left(\frac{2\pi vBd}{c} \right) dv}{\int_{v_0 - \Delta v/2}^{v_0 + \Delta v/2} |A(v)|^2 dv} \quad (4.38)$$

where v_0 is the central frequency of the line and Δv is the FWHM linewidth. Reference [10] shows that Eq. (4.38) has an analytical solution when $|A(v)|^2$ is either an Lorentz function or an Gaussian function. Here we only consider $|A(v)|^2$ is a Lorentz function since this is usually the case. Assuming $|A(v)|^2$ has a normalized form

$$|A(v)|^2 = \frac{\Delta v}{2\pi} \frac{1}{\left(\frac{\Delta v}{2} \right)^2 + (v - v_0)^2} \quad (4.39)$$

where $\Delta v/2\pi$ is a normalizing coefficient, the integral range in Eq. (4.38) can be extended to $(-\infty, +\infty)$ since for $|v - v_0| \gg \Delta v/2$, $|A(v)| = 0$, we have

$$\int_{v_0 - \Delta v/2}^{v_0 + \Delta v/2} |A(v)|^2 dv = \int_{-\infty}^{\infty} |A(v)|^2 dv$$

$$= 1 \quad (4.40)$$

and

$$p \int_{v_0 - \Delta v/2}^{v_0 + \Delta v/2} |A(v)|^2 \cos\left(\frac{2\pi vBd}{c} \right) dv$$

$$= \frac{p\Delta v}{2\pi} \int_{-\infty}^{\infty} \frac{\cos\left(\frac{2\pi Bd}{c} \gamma \right) \cos\left(\frac{2\pi Bd}{c} v_0 \right) - \sin\left(\frac{2\pi Bd}{c} \gamma \right) \sin\left(\frac{2\pi Bd}{c} v_0 \right)}{\left(\frac{\Delta v}{2} \right)^2 + \gamma^2} d\gamma \quad (4.41)$$

where $\gamma = v - v_0$. We can find from a good math handbook that

$$\int_{-\infty}^{\infty} \frac{\cos\left(\frac{2\pi Bd}{c}\gamma\right)}{\left(\frac{\Delta v}{2}\right)^2+\gamma^2}\,d\gamma = \frac{2\pi}{\Delta v}e^{-\frac{\pi Bd\Delta v}{c}} \tag{4.42}$$

and

$$\int_{-\infty}^{\infty} \frac{\sin\left(\frac{2\pi Bd}{c}\gamma\right)}{\left(\frac{\Delta v}{2}\right)^2+\gamma^2}\,d\gamma = 0 \tag{4.43}$$

since $\sin(2\pi Bd\gamma/c)$ is an odd function of γ. Combining Eqs. (4.38)–(4.43), we obtain

$$\begin{aligned} J(x) &= \frac{1+p^2}{2} + pe^{-\frac{\pi\Delta vBd}{c}}\cos\left(\frac{2\pi v_0 Bd}{c}\right) \\ &= \frac{1+p^2}{2} + pe^{-\frac{\pi Bd}{L}}\cos\left(\frac{2\pi Bd}{\lambda_0}\right) \end{aligned} \tag{4.44}$$

where $\lambda_0 = c/v_0$, $L = |\lambda_0^2/\Delta\lambda| = |c/\Delta v|$ is the coherent length of the laser, and $\Delta\lambda$ is the FWHM linewidth in term of wavelength. Term $p\exp(-\pi Bd/L)$ in Eq. (4.44) contains the linewidth information and is the key, both B and $d(x)$ can be measured and calculated with an error <10 %. Term $\cos(2\pi Bd/\lambda_0)$ in Eq. (4.44) actually does not affect the measurement result. For an analysis purpose, we assume a value for α in $d = \alpha x$ relation, when drawing Eq. (4.44).

Equation (4.44) is plotted in Fig. 4.17 for three lines with $\Delta\lambda = 0$, 0.025 and 0.1 nm, all have the same central wavelength $\lambda_0 = 1000$ nm. The corresponding coherent length is $L = \lambda_0^2/\Delta\lambda = \infty$, 40 and 10 mm, respectively. These fringes are generated by two BK7 glass flats with $n = 1.52$, $\theta = 45°$, $d = \alpha x \approx 1$ mm and 2 mm, respectively, and α is assumed to be 10″. We obtain $p = 0.96$ and $B = 3.434$ from

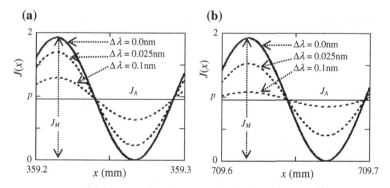

Fig. 4.17 Simulated interference fringes generated by a BK7 glass flat with 10″ wedge angle and 45° beam incident angle for three laser lines with different linewidth. **a** 1 mm thick glass flat. **b** 2 mm thick glass flat

Eq. (4.35). When performing the real measurement, the intensity of the fringe pattern is not normalized. We can measure the average intensity J_A and the modulation depth J_M, as shown in Fig. 4.17. From Eq. (4.44) we have

$$J_A = A\frac{1+p^2}{2} \qquad\qquad (4.45)$$

and

$$J_M = 2Ape^{-\frac{\pi B d}{L}} \qquad\qquad (4.46)$$

where A is a coefficient. Combining Eqs. (4.45) and (4.46) to eliminate A, and utilizing the relation $L = |\lambda_0^2/\Delta\lambda|$, we finally obtain the linewidth

$$\Delta\lambda = \left|\frac{\ln\left[\frac{4pJ_A}{(1-p^2)J_M}\right]}{\pi B d}\lambda_0^2\right| \qquad\qquad (4.47)$$

4.4 Power and Energy Measurement

Measuring the power/energy of laser diodes is not different from measuring the power/energy of other types of lasers. We only briefly summarize this topic here. The most important thing is to choose the right power/energy meter based on the emission type (pulsed or continuous), the power level and the wavelength of the laser to be measured. Laser power meters are mainly categorized by their sensor head type. The electronics used to process the data and display the results is of less importance. Coherent Inc. is one of the main vendors selling various types of laser power meters and has posted online a nice technical note about laser power meters [11].

4.4.1 Commonly Used Sensor Heads of Laser Power Meters

There are three commonly used sensor heads; photodiode, thermopile, and pyro-electric sensors. Each has its advantages and disadvantages. Tables 4.1 and 4.2 summarize their properties and their appropriate measurement tasks.

4.4.2 Integration Spheres

When we want to know the real power of a laser diode, we must measure the laser power that is directly from the laser diode without being collimated by a lens, so that the truncation loss of the collimating lens can be excluded. When we want to

Table 4.1 Three commonly used sensor heads of laser power meters are for various measurement tasks

Sensor head	Measurement type
Photodiode	Most often used for measuring low CW laser power, also used for measuring low pulse energy
Thermopile	Measures CW lasers and integrates energy of pulsed lasers to produce an average power measurement, measures energy of millisecond and longer pulse width
Pyroelectric	Only measures the energy of pulsed lasers. Average power can calculated by measuring laser repetition rate and multiplying by the pulse energy

Table 4.2 Select the right sensor heads of laser power meters based on the laser types

Laser type	Measurement needed	Power range	Wavelength range (μm)	Sensor type
CW	Average power	10 nW to 0 mW	0.25–1.8	Photodiode
		200 μW to >5 kW	0.15–12	Thermopile
Pulsed	Average power	200 μW to >5 kW	0.15–12	Thermopile
Pulsed	Energy per pulse	100 nJ to >10 J	0.15–12	Pyroelectric
Pulsed	Energy per pulse	10 pJ to 800 nJ	0.32–1.7	Photodiode
Long pulsed >1 ms	Single pulse integrated energy	1 mJ to >300 J	0.15–12	Thermopile

know the truncation loss caused by a collimating lens, we need measure the laser power with and without the collimating lens. The beam from a laser diode is highly divergent. The laser power meters described in above section often cannot catch all the laser power. Integration spheres are designed to collect the highly divergent laser power and are the best choice here.

Figure 4.18 shows the schematic of an integration sphere. The hollow spherical cavity has a diffusive internal wall and at least two windows. The reflectivity of the internal wall is high and slightly wavelength dependent. The highly divergent beam under measurement is incident into the sphere from one window. The photodetector

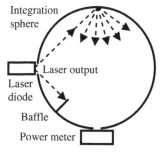

Fig. 4.18 Schematic of an integration sphere

of a laser power meter is mounted on another window. A baffle is used to prevent the photodetector being directly hit by the incident beam. The sphere can collect all the incident laser power and convert the power into a diffusive radiation proportional to the power. The laser power meter measures the radiation and displays the laser power under measurement based on the calibration data of the integration sphere.

References

1. ISO Standard 11146: Lasers and laser-related equipment—test methods for laser beam widths, divergence angles and beam propagation ratios (2005)
2. Snyder, J.J.: Fizeau wavemeter. In: Proceedings of SPIE 0288, Los Alamos Conference on Optics '81, vol. 258 (1981)
3. Faust, B., Klynning, L.: Low-cost wavemeter with a solid Fizeau interferometer and fiber-optic input. Appl. Opt. **30**, 5254–5259 (1991)
4. Reiser, C., Lopert, R.B.: Laser wavemeter with solid Fizeau wedge interferometer. Appl. Opt. **27**, 3656–3660 (1988)
5. Snyder, J.J.: Algorithm for fast digital analysis of interference fringes. Appl. Opt. **19**, 1223–1225 (1980)
6. Monchalin, J.P., et al.: Accurate laser wavelength measurement with a precision two-beam scanning Michelson interferometer. Appl. Opt. **20**, 736–757 (1981)
7. Sun, H., et al.: Monitoring of laser wavelength by the method of Young's interference. Chin. J. Laser **14**, 123–125 (1987) or Chin. Phys. Laser **14**, 145–148 (1987)
8. Sun, H., et al.: Analysis of the accuracy of the period of interference fringes measured by a photodiode array. Acta Opt. Sin. **8**, 44–50 (1988)
9. Wyant, J.C.: Multiple beam interference. http://wyant.optics.arizona.edu/MultipleBeam Interference/MultipleBeamInterference.pdf
10. Reiser, C., Esherick, P., Lopert, R.B.: Laser-linewidth measurement with a Fizeau wavemeter. Opt. Lett. **13**, 981–983 (1988)
11. Measuring Laser Power and Energy Output: Coherent technical note. http://www.google.com/url?sa=t&rct=j&q=&esrc=s&source=web&cd=1&ved=0CCcQFjAA&url=http%3A%2F%2Fwww.coherent.com%2Fdownloads%2Faboutmeasuringlaserpowerndenergyoutputfinal.pdf&ei=ZVVfVMG7K8bIsATfsYDoBA&usg=AFQjCNH8ENLiA9fQoobnI16Emymsnly6uQ